Best Wishes to Bill.
Hope you find this
interesting.

The COAL COAST

The History of Coal Mining in B.C. – 1835-1900

ERIC NEWSOME

Orca Book Publishers

Canadian Cataloguing in Publication Data

Newsome, Eric, 1925-
The coal coast

Bibliography: p.
ISBN 0-920501-25-7 (bound). — ISBN 0-920501-11-7 (pbk.)

1. Coal mines and mining — British Columbia — History. I. Title
HD9554.C23B754 338.2'724'09711 C89-091263-7

We gratefully acknowledge the financial support of the Government of British Columbia.

Publication assistance provided by the Canada Council.

Orca Book Publishers
P.O. Box 5626, Stn. B
Victoria, B.C. V8R 6S4
Canada

Cover illustration and design by Susan Fergusson.
Printed and bound in Canada by Hignell Printing Ltd.

Acknowledgements

Any author writing of the past owes a debt to achivists and librarians. To the staffs of the British Columbia Provincial Archives; the Dick Institute, Kilmarnock, Scotland; Kwantlen and Fraser Valley College Libraries and the Clearbrook Public Library, go my sincere thanks. With courtesy and wisdom, they steered my feet from pitfalls. If I have stumbled, the fault is mine.

To my parents I owe particular thanks for breaking with tradition: their sons were spared having to go 'down t' bloody pit'.

To my wife,
Irene,
for everything

Contents

"Coal brought dusty, back-breaking labour to many, wealth to a few and a certain stability to the economy. It also brought to the future province the problems of the worst labour-management struggles in its history; strikes, riots, marshal law, court proceedings and anti-Oriental legislation."

James Morton
In A Sea of Sterile Mountains

The COAL COAST

Genesis

For millions of years the swamp lay hot and humid under the changing skies. Plants fell and died where they fell; more grew and fell, and more and more, until the ground was thickly layered with woody remains. The earth stirred and the swamp gently subsided and flooded to become a shallow sea.

Slowly, slowly, over an aeon of time, the mountains crumbled. Gravity moved grains of rocks to the rivers and the rivers rolled the grains to the new sea. As the rivers met the sea they slowed and in their new lethargy dropped the particles onto the old swamp material on the bed of the sea. Sea currents swirled and played with the sand so that it was strewn thickly in some places, thinly in others. Later, mud flowed brownly in the rivers and the sea sand was covered in mud, thick in some places, thin in others. Many times the restless land and the restless sea changed levels. Many times the swamp grew and many times it was buried by sand and mud and beaches of pebbles. The surface of the Earth became layered and the mud, sand and pebbles were compressed and cemented into solid and distinct layers of rock. The vegetation of the swamp was no longer so; it had decayed, but the carbon within it remained, transformed into coal interbanded with the other rocks.

But the Earth was not done. The gigantic plates of its crust strained and slipped and plunged at their junctions one with another. Molten rock oozed beneath the ground and on the surface lava flowed. Elsewhere the rock layers were bent and broken. Then the ice came and pressed down on the land. In its coming it scoured and scraped the face of the land and in its going deposited a miscellany of rocks and clays and muds to hide the scratched and scarred surface. Over time vegetation returned to clothe the glaciated land. Only in the sea-carved cliffs could the layers of rock be seen — the pale granular sandstone, the dark and crumbly shale and the shiny black coal.

Man came. Wandering tribes roamed the face of the Earth in harmony with nature. If they saw coal in the cliffs as they passed it was of no use to them. The coal waited for another kind of man, one with other needs, who often intruded upon nature, who was often greedy and rapacious.

1. From Fur to Coal

Saturday, September 15, 1832.

"Came by steam today from London to Gravesend. Embarked on the *Ganymede* at 12 1/2 PM. Moved a few miles down the river in the evening day very fine . . ."[1] The prosaic diary entry of William Fraser Tolmie, physician and newest servant of the Company of Adventurers Trading into Hudson's Bay, gave no hint of the difficulty of the voyage about to begin. It was not to Hudson's Bay that Tolmie was bound but rather to the desolate shore of northwest North America, a land inhospitable and forbidding; savage, uncultured and rude; good only for those who traded in furs. From where the *Ganymede* breasted the muddy tide of the River Thames the crow-flight distance to her destination was five thousand miles — due west. But steering west would not do. Endless probing into the rivers and inlets of the eastern shores of both American continents had made it plain that there was no short way to the Pacific Ocean. The unbroken bulk of the Americas and the realities of the wind belts of the earth would force the two-hundred-ton sailing barque to travel three times the straight line distance before her anchor buried itself in the mud of the Columbia River.

On Sunday the journey began in earnest. From the neat green fields of the English coastline the ship sailed south to where the purple hills of the Cape Verde Isles broke the monotonous seascape off the African coast. There was endless time to waste, but not for Tolmie; he passed his time profitably with studies in mathematics and geography. The ship next headed out across the equatorial Atlantic to the South American shore while Dr. Tolmie added French and German to his curriculum. Then the barque paralleled the South American shore and ran southerly for Cape Horn. Tolmie dipped into theology and anthropology, brushed-up his knowledge of medicine, and, between times, practiced the flute. Rounding the Horn in the bitter cold of a westerly gale, the ship entered the Pacific Ocean and turned to a northerly course. As the tiny, tossing *Ganymede* furrowed the waters of the South Pacific Ocean, Dr. Tolmie noted in his diary that this was his twenty-first Sunday at sea and also his twenty-first birthday. There was no hint of celebration in his words but only a personal sense of dissatisfaction: "would that I could shake off the follies of youth in thought, word and deed. . ." He was a staid and modest young Scot who grievously underestimated his own ability. The life in the new ocean was attractive; Tolmie dissected sea creatures, examined them with a calm detachment and recorded his observations with precision and in meticulous detail. The sailing barque captain, straining to get the maximum help from the trade winds, stood away from the mainland and steered for the Sandwich Islands. Beyond the tropics, the ship turned for North America and then the prevailing westerly winds scudded the *Ganymede* along the final ocean leg of her journey. At the entrance bar the the ship paused briefly and then made her slow way up the Columbia River to where Fort Vancouver stood in command of the western fur trade. When the anchor was dropped the dried Thames mud on its flukes was six months old.

Dr. Tolmie had arrived in a clean, green world fringed by snow-white peaks. What he saw was, in part, illusion; he had come to a changing world and the romantic era of the fur trade was almost over. The Oregon wilderness was ringing with the sound of axes and smoke was rising from forest clearings. Settlers were walking the trails blazed by the solitude-loving fur traders and villages were growing where lonely forts once stood. Civilization was marching west and Dr. Tolmie was destined to play his part in that march. Somewhat distant from Fort Vancouver, he would provide an initial impulse leading to the strengthening of an infant colony which in turn would buttress the British claim to a part of the northwest coast of North America. But that was yet to come; at the time of his arrival on the coast there was no such colony.

There was no doubt that William Fraser Tolmie was bright and observant. Diary entries showed that he missed little of the vegetation and animal life or of the form of the earth and its rocks. Keen

Dr. William Fraser Tolmie, from his mainland post of Fort McLoughlin, reported rumors of coal on northern Vancouver Island.

observers were valued by the Company; that was made clear when an overland trip was frequently delayed by his companion's frequent darts into the bush to prospect for coal. The Hudson's Bay Company was on the coast for profit and there was no profit in keeping Tolmie at Fort Vancouver — he might be a doctor but he was a fur trader first. Fort Vancouver had a doctor — Dr. John McLoughlin, a fur trader of distinction, famed for building a string of coastal forts reaching up to Alaska. One of the forts had a vacancy.

Dr. Tolmie was shipped north and just before Christmas Day in the year 1833, disembarked from the *Cadboro* to take up his duties at Fort McLoughlin, far up the mainland coast beyond the northen tip of Vancouver Island. There he learned of coal deposits on Vancouver Island and set in motion a chain of events that influenced west coast development for the next hundred years.

The story goes that an Indian visiting the blacksmith's shop at Fort McLoughlin became interested in the coal being used on the forge and asked where it came from. He seemed amused when told that it came from halfway round the world by ship. There was reason to smile; the ground around his village was littered with a similar type of black stone. Dr. Tolmie on being told of coal near at hand on Vancouver Island passed the information to Dr. McLoughlin at his Fort

Vancouver headquarters and early in the following year the steamship *Beaver*, newly arrived on the coast, was sent to check the site. According to John Dunn, who was on the *Beaver*, "Mr. Finlayson . . . after some enquiries and a small distribution of rewards, found from the natives that the original account given at Fort McLoughlin was true. The coal turned out to be of excellent quality, running in extensive fields, and even in clumpy mounds, and most easily worked all along that part of the country."[2] The story is neat, traditionally accepted, and possibly true. Confirmation cannot be made from Tolmie's diary for throughout the year of the coal find his entries were sparse. For example, there were no entries at all from October 25 until December 23 when he started to write again with the observation, "Since last entry indolence has prevented me from journalizing although several events worthy of notice have occurred." Was the report of coal on Vancouver Island one of the events worthy of notice? Was the discovery as accidental as it seems? In future years Tolmie was not slow to claim credit for the find and to insist that throughout his time at Fort McLoughlin he had deliberately encouraged the natives to search for coal.

So the Hudson's Bay Company knew of coal. But what was to be done with it? The white population, what there was of it, occupied a handful of scattered trading posts and had use for only small quantities of fuel for the forges — no more than could be conveniently shipped in from Britain along with other cargo. The steamship *Beaver* could burn coal but normally relied on wood, supplies of which could be found everywhere along the coast. Whatever other infrequent ships made passage along the coast could be guaranteed to be under sail. The coal lay at the newly named Beaver Harbour as it had always done — still useless. But the Company would not likely forget a potential source of profit.

When news of the the Vancouver Island coal find was passed to the Admiralty in London, coal was of little use to the British Navy. The Lords of the Admiralty were waging a fierce rear-guard action against the need to use coal in steamships rather than using the clean, free wind in ships of sail. Such a change was earth-shattering to contemplate. A general acceptance of steam ships would fly in the face of hundreds of years of tradition, would bring about the need to replace the entire fleet, would make useless the training and experience of thousands of sailors, would force the creation of new tactics, would, in fact, destroy the world in which the admirals themselves had established their competence. They must inevitably lose the battle; the grimy, smoke-trailing steamers would take the wind out of the sails of the Navy. But not yet. The Admiralty, like the Hudson's Bay Company, also had a collective memory for information that could be of future value: the coal was of no use to them at the time but it might be of importance some day.

In August, 1841, two men occupied a room in the main building of Fort Vancouver. The atmosphere was strained, for a common loyalty to the Hudson's Bay Company did nothing to lessen their dislike of each other. Dr. John McLoughlin, Chief Factor of the Company's Columbia division, sat on a chair that seemed overwhelmed by his great bulk. His head was averted from his companion and his intense blue eyes gazed unseeingly at the scene outside a nearby window. Untidy white hair framed his morose face. George Simpson, Governor of the Company's territory from Hudson's Bay to the Pacific Ocean, studying the craggy profile, was reminded of his first impression of John McLoughlin — like a highwayman of old and a man he would not care to meet in a dark lane at night.[3] The old man's strengths were beyond doubt, but Simpson knew he had a violent temper and could be a disagreeable man with whom to do business. Yet business had to be done.

Simpson spoke rapidly, with precision, force and the assurance of one who is guaranteed the last word. He knew his facts. The fur trade was no longer profitable on the northwest coast. The Company operated sawmills, traded in salmon with the Indians and ran farms producing food to be sold to the Russians based in Alaska. Wool and hides from the Company's farms were being exported to England. He saw no justification for McLoughlin's chain of northern forts, existing to trade only in a dwindling supply of furs for a market that had sadly declined. The expense of staffing the forts was ruinous and it would be better by far to close some of them and send the *Beaver* up the coast at regular intervals to pick up what furs were available. Nothing pleased Simpson. Even the location of Fort Vancouver was questionable. Of what use, he asked, was a centre of trade lying so far from the ocean? The upriver journey could take up to three wasted weeks and the wait for conditions safe enough to allow a crossing of the bar of the

Columbia River interminable. The main depot of the Columbia division should be on the coast so that a proper and economical use of the Company's ships could be made. There must be changes.

On the outward journey Simpson's ship was delayed three weeks at the Columbia bar; an apt illustration of his point about the unsuitability of inland Fort Vancouver as a trading centre. Simpson, the realist, had read the economic and political signs and to the London headquarters of the Company he sent suggestions for re-organization. Dr. John McLoughlin's empire was about to crumble. One year after Simpson's visit, McLoughlin was told to give up three of his northern forts; in future the *Beaver* would steam up the coast and collect whatever furs were available. There was an additional order: a depot was to be set up on the southern shore of Vancouver Island and it would be the new trading centre for the northwest coast.[4] Vancouver Island, so far neglected, was about to join the world of commerce.

There was reason for haste. American settlers were flocking into the Oregon Territory. They saw the Company as an intruder and an obstruction to progress that must be speedily removed. The time was not far off when the Company, although maintaining some rights, would be effectively denied the use of Fort Vancouver.

Chief Factor James Douglas, sent to carry out an exploration of the southern coast of Vancouver Island, made haste and in July, 1842, reported that a site for the new depot had been selected.[5] He had found no shortage of harbours along the coast but, with one exception, all were closely pressed by rocks and forests and had poor agricultural prospects. Although its harbour was no better than many others, the exception was set in a plain six miles square with good farmland. There was a sufficient supply of timber locally available for construction of the fort which must be built. In March of the next year Douglas returned and began to build. The fort, first called Camosun and then Fort Albert in honour of the Queen's Consort, by a later resolution of the Northern Department of the Company came to be known as Fort Victoria. By October the stockade and bastions were finished, Charles Ross was left in charge, and the Hudson's Bay Company had a footing on Vancouver Island.

Was a footing enough? In 1846 the Oregon Treaty gave to Britain the mainland territories north of the 49th parallel of latitude and the whole of Vancouver Island. All land to the south became the property

A stiff man in a stiff pose, James Douglas was tenacious in his determination to find coal of economic value.

of the United States. The British government had learned a costly lesson: no longer was it possible to hold territory by claims based solely on discovery or by marginal occupation for trading purposes. Demonstrably, a new principle of gaining sovereignty had been established that recognized the validity of settlers occupying "empty" territory and then calling for the mother country to extend both her protection and her laws to them.[6] A footing had not been enough to enable Britain to hold the Oregon Territory.

It could happen again. The existing boundary was of very little importance when the side of the line said to belong to Britain was occupied only by a handful of fur traders. Was it likely that a line drawn on a map would deter irregular settlement leading to a future loss of territory? Earl Grey, the British Colonial Secretary of the time, thought not. It was his intention to strengthen the British claim by setting up a Pacific Colony with its own local government acting within limitations set by the home Parliament in London. To do so would be difficult. How could British emigrants be persuaded to make the long and dangerous sea journey? What would they do in the colony? So far removed from the normal routes of commerce, where would they find markets for whatever goods they could manage to produce? Inevitably there would be a cost to consider in setting up the colony. Not less than fifty thousand pounds, thought Earl Grey, and with annual grants to follow, no doubt. The government could ill-afford so much. Possibly the only hope of planting a modest colony lay in entrusting it to a private company in return for trading privileges.

None of the companies making proposals to the Colonial Office were considered to have much chance of financial success on the west coast of North America. Probably they would all be wrecked on the same rock — that of the Hudson's Bay Company rights already established. In fact, with trading posts in place; with a certain knowledge of the territory; with a constant movement of ships too and from the coast, and with its long standing financial stability, the Hudson's Bay Company seemed to be a logical choice to run a new colony.[7] The Company was interested and in its first proposal indicated an ambitious willingness to accept a grant of land extending from Hudson's Bay to the Pacific Ocean. A later suggestion included all territory from the crest of the Rockies to the Pacific shore and this too was rejected. Earl Grey finally limited the grant to Vancouver Island alone and, after much parliamentary opposition, the grant was made to the Hudson's Bay Company. The suggestion that James Douglas, then the Chief Factor of Fort Victoria, be appointed Governor of the colony was not accepted. The government of the island colony was meant to be kept very distinct from the affairs of the Company. When it was ready the Imperial Parliament would appoint a Governor — it would be ready when Earl Grey found a suitable candidate who would take on the task without expecting payment.

Settlement and the fur trade had always been in conflict and now under the terms of the grant the Hudson's Bay Company, fur trading company without peer, was to be made responsible for settlement. A curious situation; rather like the fox being set to guard the chickens! It is possible that the Company accepted responsibility for colonization only to ensure that the influx of newcomers could be controlled to its own advantage: possibly the move of the Company to trade in articles other than furs had made the coming of settlers less of a problem. Soon the Hudson's Bay Company would need settlers of its own to develop the coal deposits of the region which Dr. Tolmie had brought to the attention of the Company a decade before.

Often sailing ships of the British Pacific Fleet had threaded a difficult passage through the confined channels inshore of Vancouver Island where strong tidal currents, rocky shores and variable and uncertain winds made for dangerous sailing. In 1848 life was made a little easier by the arrival of the *Comorant*, a naval steamer assigned to tow sailing ships through the more hazardous places. The steamer was fueled with coal brought from the mines of far off Wales in ships trading in guano on the Chilean coast; ships which brought the coal at low cost because they had no outward cargo. But there was a consideration above cost and convenience of obtaining coal supplies — events in the Oregon territory had made Vancouver Island strategically important to the British position on the west coast. With an accelerating use of steam propulsion in naval ships the time had arrived to investigate the coal deposits which had lain undisturbed since their discovery. Captain Gordon of the *Comorant* visited the Beaver Harbour deposit and what he saw convinced him that the coal was "at least equal to good Scotch coal"[8] while the Museum of Practical Geology, sent a sample to examine, reported less enthusiastically that it was suitable for steam propulsion.[9] Never slow to look after its financial position, the British Government moved to ensure that the coal be reserved for the Crown. Captain Courtnay of the *Constance* built a hut to show occupation and posted a proclamation in the Queen's name warning possible intruders against mining the coal. To add point to the claim he also urged James Douglas to have his men evict anyone settling near the coal deposits.[10] As the only people in a position to exploit the coal were the employees of the Hudson's Bay Company the letter might well have carried a barb, particularly as the negotiations for mining the coal were being carried out. The Colonial Office had no objection to the Hudson's Bay Company mining the coal providing it would pay a lease on the coal lands, pay royalties to the government for each ton of coal mined and, in addition, guarantee coal to the Royal Navy at a fixed low price. The Company with its usual commercial adroitness side-stepped the issue by piously protesting that any profits made from operating the coalfield would be spent on colonization and the Company would not gain financially.[11] Agreement was reached, in time, and the Company managed to keep free of most of the restrictive proposals, which left it free to develop the natural market for coal on the west coast of the United States.

The Company had good reason to avoid limitations on its freedom of commercial activity for a coal customer of some importance was already in sight. William Aspinwall, a ship-owner of New York, had received a contract from the United States Government to transport mail between Panama and the Oregon coast and intended to use three paddle-wheel steamers for the purpose. Learning of the Vancouver Island coal deposits, he gave up his original intention to use imported coal from Wales in favour of what seemed to him a cheaper and more convenient supply.[12] Sir George Simpson wrote to the west coast as soon as an agreement with Aspinwall was firm asking that coal be mined and stored, using Indian labour if necessary. Concerned about safety he suggested that a fort be built at the coal site and that the *Beaver* should be stationed as near as possible to the coal deposit to provide protection until the fort was completed. Simpson had already engaged skilled miners and went on to say: ". . .this month, are to be forwarded a headsman and six miners, . . ."[13]

The Company of Adventurers Trading into Hudson's Bay had become grubby west coast coal merchants.

2. Fort Rupert

Why have I been left here? What has gone wrong? The answers the pale young man could find were no more satisfactory than they had been for the past month. As he paced the the hotel room, he brushed his dark, flowing moustaches to left and to right with a nervous gesture. In the fashion of the day he was dressed in dark, heavy clothing, ideally suited to England but not to the tropics where he was stranded. He was hot and angry, and looked it.

Suddenly, mind made up, he stopped before a carved desk and threw himself on the chair before it. He rose, made as if to take off his coat and then, as if remembering who he was and to whom he intended to write, shrugged his shoulders back into the coat and sat down again. Impatiently, he selected a piece of notepaper and taking up a pen dipped it into the inkwell and began to write. The pen squeaked over the paper until he paused and checked for errors. The page was headed: Panama, December 26, 1849; the salutation: Earl Grey, Principal Secretary to the Colonial Office, London. He wrote the body of the letter in polite terms yet with an edge of aggravation clearly showing through. He has arrived in Panama, he reports, but has no notion of how his journey is to be continued. He intends to write to Admiral Hornby, Commander of the British Pacific Fleet, who was to have had a ship waiting for him in Panama a month ago, and inform him of his predicament. Then he must wait until the Admiral can provide a passage in one of Her Majesty's ships proceeding northward along the Pacific coast of North America. The writer stopped, read over what he had written and, with a flourish,

Richard Blanshard, first Governor of the Colony of Vancouver Island, found the first miners to be so discontented with their employer that they could scarcely be persuaded to work.

scrawled his signature: Richard Blanshard. He hesitated and then, for the first time ever, added the title: Governor of the Colony of Vancouver Island.[1] When he arrives at his destination, he will be a Governor without salary, with powers of government but with no colonists to govern in a colony where the land and natural resources are under the control of a private company.[2]

There was no reason for haste; a government, of sorts, was already ruling the island more effectively than he ever would. The lack of a Governor was creating no problems — governmental problems would start with his arrival.

Three and a half months later Blanshard was standing at the rail of *HMS Driver* staring gloomily at the fur fort which was to be his home. Seldom had he seen such a scene of wild desolation. He saw treed hills under a leaden sky. He saw, on a small plain, a rough stockade of logs enclosing a collection of rude buildings some of which dribbled smoke which was whipped away by the wet, gusty wind. Fort Victoria looked no more than a child's hastily constructed toy set down in an immense, lonely wilderness. A flash of bright colour as the Union Jack snapped smartly on its staff provided the only cheer. The day was too far spent to go ashore. Tomorrow, after suitable arrangements had been made, he would step ashore with the ceremony befitting one who came as the

Queen's first representative to the colony.

On March 11, 1850, Richard Blanshard was rowed ashore in one of the ship's boats. Behind him the *Driver*, dressed overall, thundered out a seventeen gun salute, the noise rolling and echoing among the hills. As the salute from the ship ended, an answering salute from one of the fort's bastions started to boom out its messsage of respect. The boat came alongside a rough landing stage and, as two sailors held secure fore and aft, the Governor stepped over the gunwale and started up the path to the fort. A stocky man, plainly and soberly dressed and with a strong grave face, stepped forward from the welcoming party with his hand extended in greeting. Richard Blanshard, first Governor of Vancouver Island, and James Douglas, Chief Factor of the Hudson's Bay Company fort at Victoria, the man originally suggested for the post of Governor, solemnly shook hands. There was a flurry of introductions and the ceremony began.

The naval officers and sailors, in best shore-going rig, were lined up with their backs to the ship. Facing them stood an assortment of Hudson's Bay employees dressed in clothing from the formal to the eccentric but with even the worst dressed having some sartorial touch signifying the importance of the day. Standing between the two groups, Richard Blanshard read his commission to the assembly. It was a brief event but an important one for it marked the beginning of formal British rule on the western shore of North America.[3]

After the ceremony Governor Blanshard discovered that, important though he might believe himself to be, no one had prepared any accommodation for him. Having no alternative the new Governor returned to the Driver and, after visiting Fort Nisqually in the United States, sailed north along the coast of Vancouver Island to Fort Rupert. Fort Victoria was to be the centre of government and was already the centre of trade: Fort Rupert, in Blanshard's own words, was "for the purpose of working coals with which the North Eastern part of the island is said to abound."[4]

That the Muir family was leaving for America probably caused no very great stir in the Scottish mining region around Kilmarnock. They were no more than a fleck of foam in an ocean of immigrants bound for American shores. From 1770 to 1890 almost eleven million people from the British Isles would make the same journey[5] and in 1850 alone, two years after the Muirs left Scotland, a quarter of a million would leave from the single port of Liverpool, tightly packed in almost six hundred ships.[6] Emigration was no novelty in Victorian times, but the Muir's departure had some novel elements. Unlike most emigrants they were bound not for the populous eastern seaboard of the United States, but for the far northwest coast to a land described even several years after their journey as "a region of the globe inhospitable beyond description . . . having no title to be reclaimed by industry or visited with the benefits of civilization."[7] Surprising also was their employer and the apparent security of their move. Of all things they were to dig coal for a fur trading company and had signed on with the Hudson's Bay Company for a three-year period at a known and guaranteed wage. How lucky they must have thought themselves in comparison to the vast bulk of emigrants who hardly knew where they would go in North America or where they would find work. In addition they had the certainty of a paid passage home at the end of the contract if they decided not to stay.

The Company in its search for miners had chosen what was essentially a family unit. John Muir (Sr.) was to lead the party to Vancouver Island and be Oversman at the Fort Rupert Mine. As miners he would have his four sons; John, Andrew, Robert and Michael; his two nephews, Archibald Muir and John McGregor; and also John Smith. Michael Muir, listed as a miner, was nine years old, a common enough age for entry to the Scottish coal mines of that day. Having an obligation to promote settlement the Company had made sure that the party would include women and children. John Muir (Sr.) was to be accompanied by his wife, Anne, as well as by his widowed daughter, Marion, and her two children. John McGregor would take his wife and her sister, as well as the sister's three children. John Smith also had a family to take. Not only would the women and children help to show that the Company was taking seriously its duty to promote settlement but they would also anchor the men and make it difficult for them to suddenly depart for the goldfields of California.

Little would be known of the voyage of the Muirs and of subsequent events had not Andrew Muir kept a diary.[8] He wrote, "We started from Troon in the Royal Consort steamer to Fleetwood, should have sailed at Eight O'Clock but we were in a dreadful suspense as it was 12 Ere we saile and a very coarse night it was, we were all sick." From Fleetwood they travelled to Manchester and after a delay to change

John Muir (Sr.), first mining oversman at both Fort Rupert and Nanaimo, in later life.

their money into gold, went on to London where they arrived too late to board the ship. After a night in the Jamaica Tavern, they boarded the sailing barque *Harpooner* on the morning of November 29, 1848. There were delays and the ship had difficulty clearing the Thames Estuary but eventually the open sea was reached. A six-month journey lay before them. Informative as the diary is about the voyage, it is even more informative about the character and attitudes of Andrew Muir; factors which would play a large part in later events on Vancouver Island. As was no doubt natural on a long sailing ship voyage in the mid-nineteenth century, food was of major concern. There were few opportunities to take on fresh supplies and no efficient way of preserving such food as was already onboard. Perhaps that is why Andrew Muir gave so much diary space to conflicts between the miners and the ship's officers about the quantity and quality of the food provided. He started with praise, "pork for dinner and the pork is excellent" and, on Christmas Day, wrote with satisfaction, "a fine dinner of fresh mutton and potatoes, rum and duff." But only one day later he stormed "left the cabin after breakfast having learned they we(re) grudging the meat. I gave them to understand that I was not a tool in their hands to be used as they thought proper . . ." Later, after an argument with the ship's doctor and captain over the quality of the food, he complained that they do not know the term justice and went on to state that he hoped "one day to see justice done so that the aggressor may be punished." Another quarrel over food had him expressing the opinion that the first mate was a man of no principle. No doubt there was some justification for Andrew Muir's many complaints, but the manner of their making showed a prickly touchiness about what he considered to be his rights; an almost automatic resistance to authority; a rapidity to impute unpleasant motives to others and also a tendency to see himself as some kind of avenging angel visiting punishment upon the guilty. He never seemed to consider that he could be wrong in the slightest degree and his opinions were always expressed in extreme terms. Yet he showed little sympathy for the ship's carpenter put in irons and limited to bread and water for disobeying orders, his diary comment being, "them that's bound must obey, Servants must be subject to their masters."

Servants being subject to their masters, and lacking a ship to carry them north to the coal deposit at Fort Rupert, the miners were put to work on arrival at Fort Victoria. It was not a habit of the Hudson's Bay Company to pay men to sit idle; on the *Harpooner* they had worked their passage and they would work while waiting for a ship to carry them onward. For three months they waited and worked on such tasks as blasting rock at a dry dock site and digging wells. As experienced miners they were not pleased at this waste of their skills, but clearly there was no coal to be dug at Fort Victoria and they had to make the best of it. Finally the *Mary Dare* sailed into harbour to take them north. When they eventually did arrive at Fort Rupert, it was almost a year after leaving Scotland.

The Muir party had been introduced to life in a Company fort during the enforced stay in Victoria but that introduction can hardly have prepared its members for what they found on the shore of Beaver Harbour. As the harbour was an unsafe anchorage the *Mary Dare* stayed well off-shore and the final leg of the journey was by ship's boat which, as it neared the shore, passed through an armada of Indian canoes. The beach ahead of them was of white clamshell and at the head of a ramp sloping upward from the beach rose the white-washed wooden walls of Fort Rupert. The walls were eighteen feet high and at two corners of the square enclosure there were bastions well supplied with rifle slots and ports from which the muzzles of cannons poked. All round the clearing the forest pressed, dense and luxuriant and on what land had been cleared Indian longhouses squatted within a stone's throw of the stockade. Inside the outer gate a narrow entrance alley formed by the walls of two trading houses cut down the possibility of a concerted attack on the fort. At the far end of the alley hung another heavy gate opening onto an interior compound in the centre of which stood the houses intended for the miners and their families. The houses were rough and unfinished and probably along the lines described by the fort physician when speaking of the quarters he shared with another officer of the fort:

> . . .a round log house, the interstices between clayed, about a dozen feet square, minus a window. Inside, there was a bunk on each side, an altar in the centre, upon which a fire of wood burned; the floor, Dame Nature, covered with a few inches of sea-broken, sea-washed white clam shells. The roof had a hole in the centre, which served the purpose of letting out smoke and admitting air, light and rain — the latter pretty frequently.[9]

Inside the fort of eighty yards square, with high walls cutting off the view of almost everything but the sky, were housed twenty Englishmen, an equal number of Canadians, ten Kanakas, and the Scottish miners with their collection of wives and children. The diverse interests, attitudes and traditions of the inhabitants gave a mixture so unstable as to make conflict inevitable. All the more so because they were penned in the confined quarters of the fort by Indians given to displaying the severed heads of their conquered enemies.

The commander of the fort was Captain William Henry McNeill who had been captain of the *Beaver* when it carried the exploration party to check on Tolmie's report of coal. McNeill had made his first appearance on the coast as the owner and master of the sailing ship *Lama*, which he later sold to the Hudson's Bay Company while retaining command as an employee of the Company. He had left the sea to become a Chief Trader, but he had not left behind the training and habits of his earlier sea-going days. As an experienced sea captain of the first half of the nineteenth century, it could hardly be expected that McNeill would be highly skilled in discussion and diplomacy;

Capt. William Henry McNeill, commander of Fort Rupert and one-time sailing ship master, applied ship's discipline ashore.

more likely his orders were given to be obeyed in full and immediately, with swift, straight-forward correction meted out to laggards. Dr. Helmcken, the resident physician, saw him as a kindly man but impetuous and one who applied ship's discipline to the fort. While Mr. Blenkinsop, who was left in charge during business absences of McNeill, was described by Helmcken as good-natured, as was the next in-line, Charley Beardmore, who was also tagged as being a little "harum-scarum." The men of the "Bay" had a long tradition of self-reliance. In lands far from the easy facilities of civilization they had, of necessity, become adept at turning their hands to anything that was to be done, knowing that otherwise nothing would ever be achieved. The officers, generally, had a loyalty to the Company that was remarkable; there was no question that the Company and its interests came first. Living always in places removed from effective government, they became the government and the appliers of law as they saw it. They were in charge and if punishment was to be meted out they did not shrink from the task.

The employees of the Hudson's Bay Company at the lower levels were rather different, having little reason to be loyal to a company that worked them hard, gave them primitive living conditions and paid them very low wages. Many had come to the coast without a full realization of what they could expect and once there were trapped. No jobs existed other than those offered by the Company and there was no easy exit to pleasanter places because the departing ships, if not owned by the Company, were almost invariably about its business. The Company also found problems in the situation for, always in urgent need of labour, it had to employ whoever was available without the luxury of being selective. The journal of Fort Rupert described the workmen assembled as ". . . such a miserable set of devils I firmly believe were never before congregated . . ."[10] If they had an ambition at all it was to be somewhere else and that "somewhere," according to Dr. Helmcken, was California, where fortunes could be made with ease and lovely women, reckless pleasure and beer could be found in abundance.

If the men of the Hudson's Bay Company were a breed apart, the history and traditions of the Scottish miners had made them no less so. Scottish mine-owners had succeeded in binding their employees to them in a system of hereditary serfdom close to outright slavery. Sons had followed fathers into the mines without choice, and, so closely were they bound that, except with the unlikely permission of the mine-owner, they had little chance of ever leaving the village in which they were born. Their dismal situation caused them to be outcasts among other workers and they were so ill-regarded that the daughters of colliers were not considered suitable to become the wives of workers in other industries. In at least one area of Scotland coal miners were not allowed to be buried in the graveyard near free men and, in sentencing criminals, judges sometimes used the option of sending them to the coal mines for life. Not until fifty years before the Muirs left Scotland for Vancouver Island was serfdom in the mines brought to an end, for reasons partly humanitarian but mainly economic. Only when the increased demand for coal, brought about by the advance of the Industrial Revolution, created a demand for more miners did the old system really break down, for new miners were not likely to be attracted to work that was going to cause a loss of personal freedom. But if the abolition of serfdom was expected to draw large numbers of men to the pits, it failed; few men came forward to do a job which remained dirty and dangerous. What did happen was that many of the freed men left the mines and the chronic shortage of workers was perpetuated to the benefit of the miners who stayed. Within the lifetime of the elder Muir, and to some extent within the lives of the sons, labour scarcity enabled the miners to build up their power. After so long being considered industrial outcasts, they came to regard themselves as highly skilled workers. Then by choice sons followed fathers into the mines, there to pass through a long practical apprenticeship until they were able to take their places as hewers at the coalface. Often in later times the miners were not wage labourers but contractors on piecework employing others to take the coal from the coalface to the pit bottom. The system of piecework was an assurance to the bosses that the work of mining would be carried on as fast as possible and managerial supervision lessened. It was not long before the Scottish miners came to resent supervision of any kind and in some mines they would stop working if an oversman came round and not start working again until he had gone. The overall result was that miners came to believe that they had the right to control their own workplace.[11] Eight years before the Muirs left Scotland, a Scottish miner, Thomas Stewart, summed up the attitude of the miners when he wrote, "The principal trait of the Scottish miner is unyielding independence."[12] The Muirs' background of mining independence was a poor preparation for the treatment they received from the Hudson's Bay managers. When Governor Richard Blanshard arrived at Fort Rupert early in April, 1850, he found them fuming with frustration. They had not found coal, but for some reason Blanshard thought that they would — if they kept working. But would they keep working? Blanshard reported that they were "unprovided with proper instruments, discontented with their employers and scarcely can be induced to work."[13] In turn, the miners saw the officers of the Company as being autocratic, demanding, unreasonable and completely lacking in understanding of the difficulties of mining and the equipment needed to do the task properly.

But that was all within the fort. Dr. Helmcken saw the total situation as being, "Dangers within; danger outside; danger all round."[14] Three thousand Indians surrounded the fort, a mixture of tribes quarreling as to precedence and united only by their common dislike of what Fort Rupert meant. In the past traders had wanted furs and the Indians had wanted trade goods. Hostility was of little use in getting each what they desired and the traders and Indians had lived together in relative peace. The white man wanted only a small piece of land on which to build his trading post and another small area on which to grow food and those needs could easily be accommodated. Fort Rupert was something new. With the digging of coal the Indians had become irrelevant. The fur trade had declined and they had little hope of getting the trade goods upon which they had come to depend, for now they had nothing to offer in trade. The fur trader now coveted coal and the working of coal meant trouble, for the land from which the coal was taken was Indian land; land to which the white man had no right.

Even beyond the confines of Fort Rupert, standing in its sea of Indians, other hands were ready to add a spice of back-biting to the seething cauldron of discontent. When trouble did eventually erupt, James Douglas had words of blame for Governor Blanshard. The Governor had met the miners and "was over-familiar with them . . . and supported all their demands. Calling at Muir's house and remaining in a secret confab with the miners, was neither decorous, in a person of his rank, nor doing justice to the Company . . ."[15] That Douglas should have considered Blanshard's hearing the miner's side of the story to constitute an injustice to the Company may be the best possible indication of the attitudes making conflict certain.

The dawning day lacked substance as it broke over the leaden waters of Beaver Harbour. Patches of mist twisted and writhed, giving an amorphous quality to familiar outlines. The unreality of the dawn was echoed in the mind of Andrew Muir as it strained for solid reality and slipped back into uncertainty and disbelief. With John McGregor at his side, he had turned and tossed all night in a fruitless effort to sleep. He was weary, sore in every muscle, chilled to the bone and, above all, immersed in a feeling of helplessness deeper than he had ever known. For insisting on his rights, as every Britisher should, he had been imprisoned in one of the Fort Rupert bastions and chained as though he were a dog. Imprisoned and chained, and promised he should remain so for two years on a diet of bread and water. The new day would bring warmth, but it would not likely bring the repentance of Andrew Muir. What was there to repent? Mentally he rehearsed the events which had brought him to imprisonment. As he did so he became angry, angrier than he had ever been; angry with the blazing anger of the oppressed righteous.

The whole thing, he thought, was wrong from the start. They had been persuaded to leave a homeland rough enough but at least

Fort Rupert and the bastion in which Andrew Muir was imprisoned on a diet of bread and water for daring to oppose Hudson's Bay Company officers.

civilized, for a land rough and uncivilized beyond belief. Lured from a land settled and tame, they had been dropped on a shore wild, rugged and untamed. From houses poor but permanent they had moved into what were no better than hastily constructed hovels. They had left the freedom to roam at will over the open heather moors to be cooped up in a log fort surrounded by a horde of savages.

There had been nothing good to say about the coal-owners of Scotland but their new bosses, the fur traders, were worse. In Scotland the owners knew coal, had coal, and employed skilled miners to hew the coal: here the bosses only knew that the Company wanted coal and, therefore, must have it. They didn't even have enough sense to know that there was no coal to be had. When they had stepped ashore at Fort Rupert after being the best part of a year on the way from Scotland, they had all been anxious to get at the coal again. But there was no mine. There was not even much indication that there was coal to be mined. Here and there were to be seen thin outcroppings at which the Indians had hacked with trade axes, but the coal was salt-laden by being so near the shore and worthless. If there was to be a mine, it seemed that they would have to decide where the shaft should go, and then get on with sinking it themselves. This would not do. Andrew recalled writing in his diary ". . . we came here to work coal, not to look for it and do all manner of work . . ." But what he had written in his diary and what they must do had turned out to be different things. They had been told, simply and plainly, that they would do anything to which they were put, and, having no alternative, they did — but not quietly or with a will. The Company, without a doubt, had broken their agreement.

The fort had been built for defence against the Indians, but as there was no coal inside the fort it was obvious that the miners' work would have to be carried on outside its protective walls. The site chosen for the shaft sinking was a good distance from the fort, and they had worked amid an ever-present crowd of theatening Indians who complained loudly because the shaft was being dug to remove "their" coal. They had threatened to shoot the miners if they were not given a share of the coal when it was found. When the men had asked the Company to provide a simple picket fence around the site to give some small protection, they had been refused. And so it had continued through the long and miserable winter with the only bright spot the two shillings and sixpence per day pay increase they had demanded as compensation for the kind of work they were having to do. The threat of a strike had soon brought the Company to grant the increase so that the work of exploration would go on.

It was not long after the visit of Governor Blanshard — a fine gentleman who seemed to understand their problems and at the same time showed very little respect for the Hudson's Bay Company and its men — that the trouble really started. By the middle of April the men had grown tired of being "without assistance and protection" and downed tools in protest for a week. On returning to work they were first set to blasting and then, almost immediately stopped, and told to dig a drainage ditch. As the men saw it, there was no end to the tasks they, skilled miners, were expected to do and asked Mr. Blenkinsop, in charge in the absence of Captain McNeill, for "a paper of his hand that we were to be labourers for the rest of our time." Blenkinsop had blustered and finally threatened, refusing to provide such a paper and swearing instead to take sword and pistol to them. Andrew Muir remembered that he had been too sick to work on the next day and on returning the day after had been berated by Blenkinsop who accused him of neglecting his duty and imposed on him a fine of fifty pounds. He had gone on to say that the men spent most of their time sitting idle while he, Andrew Muir, was "a rebellious person and kept the men off their duty." This "defamation of character" had caused the men to stop work and they could not be persuaded to start again, saying that they would await the return of Captain McNeill and have the matter settled.

When, on the second day of May, McNeill returned the men were still idle and he had the problem to face and settle without delay. Andrew Muir thought he would remember the "settlement" for a long time. McNeill had called the men into the main hall of the fort and there had behaved "like a madman swearing and threatening." They had been told to get back to work immediately. On stepping forward to say that the men would not return to work until the matter had been "fairly tried by English laws," Andrew Muir was clapped in irons and imprisoned, and along with him went John McGregor. He reflected bitterly on their treatment. On the first day of imprisonment McGregor had not even had a roof over his head and only after passing a day in these conditions had he been transferred to the bastion. They had not been provided with blankets and their health had been endangered. He, himself, had caught a cold which made him partially deaf, but the doctor had laughed when told of this. Here in a British Colony, in an Empire in which slavery had been abolished, he knew "it reigns here in full force. They will always have slaves who will crawl at their feet even when they look." His bitterness was boundless. When would it end?

There was no clear cut, decisive end: the affair limped to a conclusion. After seven days in irons Muir and McGregor were taken to the main hall of the fort and offered their freedom if they would return to work. They refused to do so on the grounds that they had been wrongly imprisoned and would await trial and final settlement when the *Beaver* next called. Given time to reconsider, they still refused and were returned to the bastion but free of chains and with the promise of proper food. McNeill had created an embarrassment from which he could not easily escape and the miners were not going to make it easy for him to do so. Within two days the prisoners were freed from the bastion, but remained confined to the fort, a restriction they seem to have evaded with ease. When the *Beaver* arrived it carried a letter from Governor Blanshard appointing Dr. Helmcken, the fort doctor, a Justice of the Peace. Muir's diary entry that we are "to get

Dr. John Sebastian Helmcken, Fort Rupert physician, was thought by Andrew Muir to be the man able to redress the miners' grievances.

our grievances redressed by Dr." had little hope of becoming reality. Dr. Helmcken, himself an employee of the Hudson's Bay Company and directly under the command of the man against whom the miner's complaint was lodged, would have a very slim chance of arriving at an impartial and acceptable solution. He had no illusions about the impossibility of his task and compared it to "making a ship's doctor justice of the peace in a ship at sea! If the crew became mutinous, who was to act — mate, magistrate or captain?"[16]

Meanwhile the crew of the United States registered ship *Massachusetts*, in harbour for coaling, had been spreading unrest among the fort's inhabitants with wonderful stories of how rich it was possible to become in California. These stories fell on receptive ears and were apparently swallowed without anyone asking the ship's crew why they were not in California becoming rich. The *England* next arrived for coal. It provided both a means of escape to California and an additional reason for leaving Fort Rupert. It was known that several deserters from the Hudson's Bay ship *Norman Morrison* were onboard, but, as a search failed to find them, it seemed evident that they had somehow escaped to the shore before the search started. A reward was offered for the return of the fugitives. Soon afterwards they were found murdered. In the over-heated atmosphere of discontent raging in the fort it was not long before the murder was twisted to reflect badly on the fort's officers. Andrew Muir was adamant that the deaths were due to the fact that "Blinkinsop (sic) offered the Indians 10 blankets for each of their heads should they bring back only their heads . . ." He later commented that he had "heard from the Indians that there were 3 Men killed belonging to the ship — there is the effect of the blankets." Helmcken was rather more moderate in his assessment. He agreed that a reward of blankets had been offered, but said that it was not the intention of Blenkinsop that the men should be returned dead. The effect was the same whatever the intention, and it led to the intervention of Governor Blanshard and the British Navy who destroyed an Indian village in reprisal for the murders.

By the time much of this had happened, the miners were already gone, "determined to make for some christian place — since we could get neither rights nor privileges." Somehow arrangements had been made to board the departing *England* and, with the possible exception of John Muir (Sr.), his wife and young son, Michael, all boarded for San Francisco. Other employees were also trying to leave, the fort was in uproar and the search for coal was at a standstill. Dr. Helmcken, omitting as a probable cause of the desertions the employment practices of the Hudson's Bay Company, summed up the reasons for leaving as the general desire of the men to go to California. Of the miners he said, "They expected and wanted to be their own masters — uncontrolled."[17] For a while recriminations flew and blame was assigned. It was not enough for Blenkinsop to point to the Fort Rupert Journal which recorded Andrew Muir as having said, "Revolution was approaching, the Company's day was gone by," for in the opinion of no less person than Sir John Pelly, Governor of the Company, Blenkinsop should have communicated with the miners only through their oversman, Mr. Muir. As for McNeill putting them in irons, he could not conceive of anything they might have done being serious enough to warrant such treatment. In what Dr. Helmcken accurately termed a "miserable state of affairs" even Governor Blanshard did not escape censure, being reprimanded by the Colonial Office and in addition obliged to pay his own costs incurred on the voyage to destroy the Indian village.

This was not the end of the Fort Rupert coal venture, but it certainly was a poor beginning. The first and most important customer of the mine had instructed its agent to get rid of any stock of Fort Rupert coal providing a price equal to the purchase price could be obtained and not to bother to pick up the balance of the coal on order.[18] The price had turned out to be too high, the quality of the product too low and the certainty of supplies very doubtful. Even so, the Company was not yet ready to give up and brought out from Scotland Boyd Gilmour as Oversman of a new group of miners. In spite of all their efforts no coal of commercial value was found.

By mid-June of 1851 Governor Blanshard was reporting to the Colonial Office that the Muirs, having left Fort Rupert because of the ill-treatment and threats meted out to them by the officers of the Hudson's Bay Company which had caused them to be in fear of their lives, had returned to seek redress for the losses and damage they had sustained.[19] In fact the petition enclosed with Blanshard's letter named only Andrew Muir who laid a complaint against McNeill, George Blenkinsop and Charles Beardmore. He charged that they had imprisoned him for twelve days in irons and on a diet of bread and water which had resulted in permanent and serious damage to his health. As it was, Governor Blanshard was unable to grant redress: he was leaving Vancouver Island with grievances of his own and was to be succeeded by James Douglas who would also remain in his position with the Hudson's Bay Company.

The island's first settler, Captain Walter Colquhoun Grant, remarked of the Fort Rupert venture that both the quantity and quality of the coal had been vastly over-rated.[20] Apart from the surface coal first dug by the Indians there was no coal of commercial value. The miners employed had "not raised one square inch" of coal. Captain Grant went on to say that coal had also been reported on the east coast of the island but he suspected that it would only be surface coal as was the case at Fort Rupert. What he knew of the geology of the island suggested to him that "coal mining will never turn out to be very productive on it."[21] He was wrong.

3. Fort Nanaimo

There is a curious parallel between the events at Fort McLoughlin in 1835 and those at Fort Victoria fourteen years later. As told by Joseph McKay,[1] then a clerk employed at the fort by the Hudson's Bay Company, an Indian visiting the blacksmith's shop picked up a piece of coal and remarked that there was plenty of such stone where he lived. McKay promised a trifling reward for samples and the incident passed. In April of the next year the Indian returned with a canoe load of good quality coal and McKay was sent out with a prospecting party to examine the reported coal source. After several days of exploring a good seam was discovered and McKay returned to make a favourable report on what he had seen. At that time exploration of the Fort Rupert mining area was still going on and the Company saw no reason to exploit the new find. Only when it became clear that the Fort Rupert coal was something of a myth did James Douglas send McKay out again to re-examine his earlier find. In the late spring of 1852 he returned to tell of seams up to a foot in thickness outcropping in the cliffs and at Point Gabriola a seam measuring no less than thirty-seven inches.[2] James Douglas then carried out a personal examination, taking with him John Muir, the former oversman at Fort Rupert, and Mr. Pemberton, a surveyor. Having built a fort to guard a worthless coalfield and having greeted miners sent from the other side of the world to develop it, Douglas seemed driven to prove that there was coal of high value on Vancouver Island. There was plenty of coal to be seen on the shores of Wentuhuysen Inlet and from Victoria he wrote a letter to Archibald Barclay, Secretary at the London Headquarters of

The man who told of Nanaimo coal deposits, "Coal Tyee" (centre), dressed for the cold but barefoot.

the Company, crowing that the wealth of minerals had not been overestimated. On one side of an island in the inlet he had seen a seam of coal which showed also on the other side, suggesting a continuous seam not very far below the undisturbed surface rocks. A mainland seam three quarters of a mile away had a thickness of almost six feet and dipped towards the water in such a way as to give the impression that it might continue across the island at a deeper level than the visible coal seam. The Chief Factor by using Indian labour was able to collect fifty tons of coal for the trifling outlay of only eleven pounds sterling in trade goods. The coal, taken out on the *Cadboro*, was the first shipment of coal from the area. Already Douglas had in mind employing John Muir and some of his sons to carry on further exploration and envisioned, at some future date, withdrawing the miners still working at Fort Rupert and setting them to work on the new site.[3] For a man of unbending dignity, the open enthusiasm of James Douglas for the coal find was startling.

The Hudson's Bay men had never been laggards when there was a profit to be made, but the speed with which Douglas acted on his return from what he called "the coal district about Wentuhuysen Inlet" was astonishing. Within one week a party had left Victoria to establish a claim to the newly discovered coalfield; a surveyor was heading north to make a detailed geological survey and a ship with miners onboard had sailed with instructions that it was to return loaded with coal. Joseph McKay had instructions to "proceed with all possible diligence to Wentuhuysen Inlet commonly known as Nanymo Bay and formally take possession of the coal beds lately discovered there for and in behalf of the Hudson's Bay Company."[4] The call for haste was superfluous as McKay had left Victoria for Nanymo Bay two days before James Douglas dated his written instructions which were no doubt intended as a confirmation of verbal instructions already given. McKay was to forbid all persons to gather coal; to levy a royalty of two shillings and sixpence on each ton of coal removed by persons licensed to do so by the Company (clearly no one was licensed, or was likely to be); and he was to let Douglas know if anyone broke the regulations.[5] In short he was to make obvious the claim of the Company to the coal deposits and chase away intruders. No mention was made of the Indians who owned the land from which they, apparently, were also forbidden to remove coal by the edict of a foreign commercial company.

McKay was told that the miners being sent were John Muir, his son Robert and nephew Archibald, and also John McGregor, a fairly recent companion in chains of Andrew Muir in the bastion of Fort Rupert. There was a caution that "the Miners are under the special instructions of Mr. Muir and you will please avoid all interference with them directly, giving any instructions you have to give through Mr. Muir himself, but in no case directly to the men under his orders."[6] The Fort Rupert experience had taught James Douglas to be wary of the sensitivities of coal miners and of the Muirs in particular. Furthermore, the miners were to be given one shilling per diem for rations as that would avoid conflicts about the quantity and quality of food provided by the Company.[7] McKay had been gone only a week when Douglas wrote again to say that a second ship, the *Recovery*, was heading north and was to be returned to Victoria with a full load of coal as soon as possible. He concluded, "Please write to me at every opportunity and let me know particulars about the coal and if any fresh discoveries have been made since my visit."[8] If James Douglas had wrought miracles of speedy organization, Joseph McKay was expected to do no less.

Throughout September Joseph McKay had nothing but good news to report. The *Cadboro* had been loaded with four hundred and eighty barrels of coal by Indians working for trade goods, their rate of pay being such that "An able man can earn at the rate of one shirt per diem."[9] Mr. Muir, having found a coal seam seven feet below the surface after digging ten yards away from where the Indians obtained the coal to load the *Cadboro*, had moved on ten yards to dig again.[10] It seemed that there was a great profusion of seams and by mid-month a seam six feet two inches in thickness had been found. Muir considered the coal to be equal in quality to good English coal and was of the opinion that the natives alone might have raised ten thousand tons; McKay, who might have had to deliver any quantity that he named, cautiously calculated that a thousand tons could be raised fairly quickly. He pressed for more trade goods to pay the Indians: woollens, shirts and ammunition he must have; cut glass beads, tobacco, kettles and soap were also in demand and more would be useful. McKay had also made the first private sale of coal to Captain Webster of the *Honolulu Packet*. There was a small problem with the sale — Captain Webster had no money with which to pay. He had been permitted to take on thirty-two tons of coal, explained McKay, on the assurance that the ship must necessarily return to Victoria where the ship's chronometer and papers were being held by the Governor. Money was available to him in Victoria the Captain claimed. The bill was for three hundred and twenty dollars.[11]

By mid-October Douglas was telling McKay that Gilmour's miners would require comfortable lodgings on their arrival from Fort Rupert.[12] The decision had been taken and Nanymo Bay was to become the centre of the Hudson's Bay Company mining operations on Vancouver Island. Two weeks later James Douglas was anxiously awaiting details on a particular coal seam that was being explored. When advised of its potential worth, he intended to "address the Company on the subject of sending out an additional body of miners . . . from England."[13] The Muirs and their companions, picking and digging on the shores of Nanymo Bay, were laying the foundations for a major industry on the site where the city of

Nanaimo now stands.

There is a gap in the letters gathered together as the "Nanaimo Correspondence" which leaves blank the winter of 1852-1853, but when the letters resume in April, 1853 the Fort Rupert men had moved in and there was money trouble. The miners reckoned that if they raised the expected thirty tons of coal per man each month their pay was equivalent to four shillings each ton, but for each extra ton raised the Company was offering to pay only two shillings and sixpence. To the miners the effort of extra production was not worth-while. In calculating the cost to the Company of each ton of coal the men had included the daily one shilling allowance for rations — perhaps the Chief Factor's instruction to McKay to give a money allowance instead of food rations, "which will save you so much trouble,"[14] had brought the very trouble he wished so earnestly to avoid. Worse, the men working as assistant miners at half the pay of miners had the idea that in consequence they should raise only half as much coal and then be paid two shillings and sixpence for each extra ton.[15] In a masterly feat of reasoning James Douglas included the bed and board cost paid to the assistant miners to show that they were, in fact, paid much more than half the pay of the miners and so should expect to raise more than half the quantity of coal, while omitting the same cost when calculating the per ton rate of the miners and making it seem lower than the miners had claimed and, therefore, the rate for extra tons produced seem more reasonable. He instructed McKay to offer the miners two shillings and ninepence for extra tons of coal. The assistant miners could have two shillings and sixpence for extra tons, but only after producing in the month three-quarters of the tonnage demanded of miners.[16] Shortly after the hard-pressed McKay made the offer, he had to report that the miners would not increase their monthly return of coal and the assistant miners had refused to go beyond the fifteen tons they had claimed to be reasonable for their rate of pay. The assistant miners had been taken off coal digging completely and were being employed around the fort on a variety of tasks.[17] Douglas replied, stiffly, that the assistants must choose to be either miners or labourers and that making an example of the leaders might bring the others into line. He confessed to being discouraged at having so many problems when two oversmen were available to superintend five miners and a half dozen labourers.[18]

James Douglas expected the arrival of "40 miners, 36 of whom are married and have altogether 87 children."[19] While serving its own interests by importing miners to develop the coal mines the Company was keeping its obligation to promote settlement in a manner that would bring no problems in its train. The community as it now existed was predominantly male, but there were some women and children. Before the population was increased by immigration, McKay had entered in his journal a "native" growth: "Two births have occurred at this establishment . . . in the cases of Mrs. Dunsmuir and the native wife of John Malcolm, labourer."[20] When the newcomers arrived there would be a deluge of women and children that would bring a maturity and balance seldom found in a pioneer settlement on a distant frontier. Joseph McKay now had a new problem; that of housing the expected settlers. Governor Douglas had commented already that "as to work and food we can provide abundantly, but I am not so sure about house accommodations."[21]

Houses and buildings there were of various shapes and conditions from the rough log and cedar bark huts of the early days to houses of a more permanent and satisfactory kind. By June, 1853 the settlement on Nanymo Bay had four houses of four hundred square feet in area, three of six hundred square feet and three more of the same size were framed and awaiting completion. A bastion had been built through the axe artistry of Leon Labine and Jean Batiste Fortier; there was a water-powered sawmill and a large store was about to be built for the Hudson's Bay Company.[22] Even before Douglas had been warned of the impending arrival of the new settlers he had engaged Francois Cote to build dwelling houses and four or five were expected to be ready for use by spring, 1854.[23] But even with the new construction the number of houses would still be insufficient to meet the coming demand and, on the recommendation of the Company, Douglas had his men set about building twenty detached houses, each one of which was to be provided with two entrance doors so that division to accommodate two families was possible.[24]

As it turned out, the rush was not as great as anticipated. There were delays and on the date the houses were to be occupied their eventual occupants were still living in the dirt and gloom of England's "Black Country."

When Satan stood on Brierley Hill
And far around he gazed,
He said, 'I never shall again
At Hell's flames be amazed'.

Traditional Ballad[25]

The Industrial Revolution had cast its grimy net over great tracts of England by the time of the Victorian "Golden Age" turning them into regions of flame, smoke, dirt and noise. Nowhere was this more evident than in the "Black Country" of the South Staffordshire coalfield. Blast furnaces, dilapidated industrial buildings, mean brick houses, dumps of factory waste and rusty metal, stables, outhouses and sheds littered the countryside. The repetitive thump of heavy steam-driven machinery, the roar of blast furnaces, the clank and squeal of primitive locomotives hauling open trucks of coal over metal rails filled the air with a harsh cacophony of sound. Here and there, gaunt pit frames stood with their flickering winding wheels starting, accelerating, slowing, stopping and then reversing endlessly as miners descended pit shafts and tubs of coal were raised. Near each pit straggled a mining village set in a bleak and unlovely landscape of black pit heaps. Man and his grubby activities predominated. Production never ceased. At night the sounds were muted but the blast furnaces burned on, their fiery glow lighting the sky and painting the ever-changing patterns of steam and smoke in menacing hues. It could have been Hell and perhaps for many it was.

The inhabitants of the "Black Country" were rough and straightforward. They had a simple, direct manner free of guile and pretence and often their natural courtesy seemed to outsiders to be grudging and ungracious. Nothing about their lives was likely to impart a sheen to their behaviour. Many of them had flocked to the new factories in search of what they had thought to be the new prosperity only to find themselves trapped in the stinking squalor of the crowded new industrial towns where sanitation was almost unknown and disease rampant. There were no guideposts left to them from the past, for all traditional expectations had been swept away in a new economic system in which the factory owners, selling goods dearly, were interested only in buying labour at the lowest possible price. The owners of the factories seldom knew their workers, the machines received better care than they, and, with very rare exceptions, the workers were destined to remain trapped in a rigid class structure in which the distribution of wealth had already been determined. All that seemed likely to happen was that the rich would become even richer while they, the poor, sank even further into a morass of poverty. The conflict between capital and labour had grown quickly and then hardened to become a dominant fact of life. The poor, isolated and helpless, struggled for a bare survival in a system they could not in the slightest degree influence. But the concentration of wealth and machinery had brought an inevitable concentration of workers and, as the mass of workers came to see that they had a common bond in adversity, it was from the floor of the vast and crowded factories that a hint of a solution was to come. Individually they might be weak, but collectively they could be strong. In the factory towns a consciousness of oppression grew and was nursed into a rudimentary union movement.[26]

The miners of the "Black Country" were not slow to learn the new ideas spawned in the factories and adapt them to their own use. So effective did they become at using collective pressure that a contemporary writer said of the Staffordshire coal miner and his high wage rate that he "glories in his power of maintaining it by . . . combinations and strikes . . . "[27] The same writer said that the dignity and independence claimed to be a characteristic of the miners seemed to him to be mainly a cover for their lack of self-control which made itself apparent by an enthusiasm to down tools at any chance of excitement or entertainment, be it a horse race, a local fair or a poaching expedition. In any event some of that supposed independence and dignity and that use of combined force to influence working conditions was about to be exported to far off Vancouver Island.

On May 10, 1854, twenty-three miners of the Brierley Hill Colliery committed themselves and their families to a giant leap in the dark, leaving only the warmth they felt for each other and the roughly proffered aid they knew they could count on in times of trouble as certainties.

The men looked at each other uncertainly. Who would sign first? Some men stared off into space; some nodded to friends as if to give encouragement and confidence. Then, with a low "Aye," one man squared his shoulders and stepped forward to the silent men at the table. There was a muttered conversation, too low pitched to be audible to the waiting men. Papers were passed and pages turned and finally there was a scratching of nibbed pens. The man straightened up and walked away from the table smiling down at the money in his hand. The spell was broken. Miners shuffled forward in turn to either scrawl an unpracticed signature or to make a mark. The roll was complete: George Baker, Thomas Jones, Joseph Webb, Richard Richardson, Daniel Dunn, Jesse Sage, John Thompson, Joseph Bevilockway, Thomas Hawkes, Edwin Gough, William Incher, John Biggs, Thomas Lowndes, Matthew Miller, George Bull, Thomas York, John Richardson, John Malpass, Richard Turner, Elijah Ganner, John Baker, William Harrison, John Meakin.[28] Solid young men with solid British names; men steeped in the traditions of the land they were about to leave.

Nineteen-year-old John Thompson sat fingering the Contract of Service on which he had just placed his "X".[29] His mark had been witnessed by George Robinson who was to be mine manager at Colville Town, the newly named settlement on Nanymo Bay, on the east coast of Vancouver Island. For the other party there appeared on the document the signature of W.G. Smith, Secretary to the Company of Adventurers Trading into Hudson's Bay. John Thompson might not have been able to read the document he held, but that was a reflection upon the society in which he had grown up rather than a measure of his intelligence. He was fully aware of the conditions under which he was now bound. He was also aware of the fifteen pounds sterling signing-on advance in his coat pocket; more money than he had ever possessed at one time in his whole life. There had been endless talk about the contract and all it meant, but all the discussion had given little idea of conditions in the place where they were to live for five years. What would life be like there? He could not imagine, but knew it would not be like the "Black Country." There in a far off land, among very few white people, opportunity would be theirs.

From its experiences at Fort Rupert the Company had learned about miners. This knowledge was reflected in the form of the contract — nothing foreseeable had been left to chance or to casual interpretation. It was quite clear that each man could be required to work "as a Working Collier, Miner, Sinker or Labourer." In no less than three places it was stressed that they could be used on tasks other than the digging of coal. Each man was to make himself useful as a Labourer on the outward and return passages, save that "he shall not be bound to go aloft." With the help of other Company servants and from supplied materials each man, on arrival, could be required to construct a wooden building and other conveniences for himself and his family. At the end of five years the miner and his family would be returned home, but if he was dismissed before that time his return fare would not be paid. Should he die his widow and children would be returned to England at no cost. No miner would be required to work more than "ten hours each lawful day": they would find that in the opinion of James Douglas, who would be their ultimate overseer on the island, that meant on three hundred and ten days each year.[30] Excepting Sundays, they would have a reasonable expectation of three other days annually on which they would not be required to work. For this working day and year they were to be paid the sum of seventy-six pounds sterling, but not in the first year for they must pay back the fifteen pounds of the signing on allowance within a year . . . and they would receive no pay until the six month journey was over and they were digging coal on Vancouver Island. The contract promised to pay an extra two shillings and sixpence for each ton of coal a man produced over a monthly minimum of forty-five tons: how were they to know that the miners already at work in Colville Town had limited their personal production to thirty tons monthly and considered two shillings and sixpence inadequate payment for extra production?[31]

John Thompson would pass from an organized, tamed and industrialized land marked everywhere by the activities of man to an almost empty wilderness where the traces of man were few. A greater contrast would be hard to imagine. He was to leave the only country he had ever known, with all its faults, for five years, with no chance of returning before that time had elapsed. He married the day before the ship sailed and embarked with a nineteen-year-old bride. It was almost the end of the year before the Thompsons looked with wonder and shocked surprise at their new home.

Local tradition has it that as John Thompson, the first of the party to disembark, set foot on the shore at Colville Town on November 27, 1854 a welcoming beam of sunlight broke through the grey clouds. If it was so, it was an omen of better times ahead and a signal that the pain, discomfort, trouble, and death of the past six months was at an end.

From the afterdeck of the *Princess Royal* First Mate Charles Gale looked out through the forest of masts and spars of the sailing ships crowding London's East India Docks. His own ship, fresh from the builder's yard, was cleaner than any ship on which he had ever sailed. The Hudson's Bay Company had spent wisely and well on her construction and around him he saw the carefully fashioned knees and beams of English oak, beneath his hand he felt the smooth solidity of a teak rail and in his nostrils was the satisfying smell of the oakum and pitch used to seal the deck planking. The ship had not yet felt the strains and stresses of an ocean voyage, the woodwork was free of salt stains, the furled sails were whole and unpatched, the taut rigging gleamed in the rays of the sun slanting down between the brick warehouses. The ship had been designed to carry unwieldy spars from Vancouver Island in addition to the usual and expected cargo of furs. The immediate problem facing Mr Gale was not to stow spars but people. In the various holds and cabins of the ship were to be stowed not only the crew and stores for the voyage, as well as the supplies needed by the Vancouver Island forts, but also over a hundred passengers. All had to be tightly packed on a small floating island the greatest measurements of which were one hundred and fifty-four feet in length, twenty-nine and a half feet in breadth and eighteen feet in depth.[32] Mr. Robinson, the mine manager, already occupied a cabin with his wife and two children, as did the schoolmaster, Mr. Clarke, and his family. Robinson's maid and three ladies going to join husbands already on the coast were quartered in the intermediate accommodation. For the rest — a woman with five children; ten Norwegian labourers and a party of over eighty persons from the coalfields of the Midlands — they must be content with the steerage accommodation below decks.[33] It would not be too bad in good weather but it would be hell in rough seas when the decks would have to be cleared and the hatches battened down. The ship, Gale reflected, would not smell as sweet at the end of the voyage as at the beginning.

A throng of people advanced slowly along the rough dockside cobblestones, they seemed uncertain — or possibly reluctant. Already this day they had left home, many of them for the first time, they had travelled by train to London, a city large beyond their wildest imaginings, and now they were to trust their lives on a ship which

The Princess Royal *carried Staffordshire miners to Vancouver Island at a high cost in children's lives.*

looked puny and unsafe. There were about a couple of dozen men, sturdy and roughly dressed, laden with boxes and rope tied bundles of all descriptions. A similar number of women trailed behind, some with packages but more carrying infants and young children while other children held tightly to their hands or dragged at their skirts. By the appearance of some of the women the ship's complement would be larger on arrival than on departure. But it was not necessarily so. Only one of the three children born on the voyage would survive. Of the wide-eyed children on the dock five would not see Vancouver Island. One miner and the wife of a miner would find a last resting place on a tropical island far from home. No sailing ship voyage of that age was easy; the voyage of the *Princess Royal* would be no different.

Mr. Gale signalled to the tug sitting in the middle of the dock basin. There was a hoarse "toot" of acknowledgment followed by the slap — slap — slap of paddle wheels rotating in the scummy water. The towrope tightened and the *Princess Royal* drew away from the dockside. As the gap widened a sailor on the stern cast off a final rope and severed the last link with land. Captain Wishart paced his deck muttering commands to his First Mate as the ships were carefully manoeuvered through the crowded confines of the dock. The eastern sky showed only faint streaks of light as the ships entered the ebbing Thames and turned seaward. Soot settled on the deck and there was a gritty feel to the teak rails of the sailing ship as she trailed after the impudent tug leading her by the nose and adding to the indignity by streaming back filthy clouds of acrid smoke. At Gravesend, where there was sailing room, the tug dropped the towrope and, with a cheery whistle blast, turned to swim splashily back upstream. One by one the sails of the *Princess Royal* were unfurled. As each was raised to catch the wind there was a slatting and an uncertain flapping before the sail gradually swelled into a firm, powerful profile. With each full sail the ship leaned a little more decisively into its task; water chuckled at the bow and then laughed out loud as the bow wave creamed higher. A groaning and creaking of masts and spars was to be heard. In the rigging there was an insistent wind song; soft and re-assuring. Until the voyage ended it was a voice that would seldom be stilled. As the *Princess Royal*, serene and dignified, settled on course Charles Gale dutifully noted in his log that at 4 o'clock on the morning of June 3, 1854 the ship left the Port of London bound for Vancouver Island. All was well.

All might have been well for the First Mate, but the steerage passengers were not so sure. On arrival the previous night they had been shown the hatch to their cavernous quarters and there left to fend for themselves. Since then they had been penned in what amounted to a bare planked box with heavy frames and sturdy deck beams cutting into it. Each family had marked out its territory with boxes and bundles making it almost impossible to move in the gloom without stepping on someone or something. Primitive beds had been made up — nests of blankets and spare clothing laid on the bare deck planks — where they had tossed and turned all night. The air below decks was stale and oppressive, men and women had snored and groaned in fitful sleep , voices had shrilled in family arguments. Women, thinking of the hard journey ahead, had softly wept. A constant whimpering and wailing of small children had provided an undertone of noise, punctuated occasionally by the sound of a sharp slap followed by a howling protest that tailed off into sobs. By four in the morning even the pretence of sleep had been ended by the scurry of heavy feet on the deck overhead and by shouted orders. There were sudden and unexplained sounds. Somewhere far outside the wooden walls a rhythmic beat increased in speed, the ship lurched and swayed and there was suddenly a different feel underfoot. Once it seemed as though the ship had hit a solid object and throughout the space there had been a booming echo and a reverberation that had set their teeth on edge. Men had rushed to the companionway to see what was happening only to be curtly ordered to stay below. The only news of progress had come from the few who were able to poke their heads out of the hatch and pass down information. There was little sensation of movement except a slight lean as the ship turned to keep to the middle of the twisting channel. Later there was a scurrying of feet followed by a squeaking of ropes passing through stiff new pulleys and as the sails were raised the deck tilted alarmingly under the press of wind on canvas. Belongings slipped down the canted deck and in the half-light people exchanged fearful glances and looked towards the oblong of sky at the top of the companionway as though gauging their chances of escape if the ship should overturn. As the river broadened the deck beneath them started to sway and pitch, at first imperceptibly and then with a firmer rise and fall. There were new creaks and groans as the ship took new stresses and strains and all too clearly the gurgle of water sounded on the other side of the oaken hull planks and on that sound was superimposed the slap of waves. They were frightened and very uncomfortable. Many wished they were safely back home in Staffordshire.

The first few days were a time of settling in. The ship got used to the open sea and the crew got used to the ship. The passengers, as people invariably will, got used to their new surroundings. Leaning to one side or the other became so common-place as to seem natural. The ship had beaten round the North Foreland, passing from the mudbank littered waters of the Thames Estuary into the clearer but confined channel between England and the dim, low coast of France. Most of the passengers had suffered in the short channel chop to the point where a sudden sinking of the ship would have seemed a merciful end to a misery too great to bear. The stench of vomit permeated the steerage space encouraging only more sickness. Gradually the passengers had slipped into a state of exhausted apathy until the desire for a speedy end slowly waned and living had to go on. The men worked around the ship as their contract demanded and gradually came to ignore the motion of the ship. For Captain Wishart and his First Mate this was the worst part of the voyage. They had beneath them a ship new and untried, the sailing limits of which they had yet to discover. The newly formed crew had yet to be tested for competence and trustworthiness. The waters in which they sailed were the most crowded to be found anywhere on the world's sea lanes. Tacking into the prevailing south-westerlies, there was always the danger of being driven on to the lee shore to starboard; a shore progressively craggy and dangerous with westerly advance. Not until they were clear of the Lizard and coastal rocks with names evocative of dread — the Manacles, Dolor Point and Black Head, the Balk and the Quadrant, the Lion's Den and the Devil's Frying Pan — would they be able to angle to the south to miss the protruding thumb of France, marked by the Ushant Light, and head for the comparative safety of the open ocean. Then it would soon be time for Mr. Gale to begin his dreary catalogue of death.

On July 1 Mr. Gale entered in his log the first death of the voyage. The baby of the schoolmaster's wife died and as the Captain spoke the traditional words of the burial at sea the tiny, wrapped body was placed over the side of the ship to disappear into the lonely immensity of the ocean. From that time Mr. Gale would record at least one death for each month of the voyage. In August the child of miner George Bull died and entered the ocean. In the month of September a child of John Malpass was buried at sea. Four days later the wife of John Baker gave birth to a daughter and, when it became clear that the sickly child would die, the parents called upon the captain to perform the christening. Next day the baby, named Ann Marie, was put to rest in the sea. In the meantime the Sage family had grown by one and this lucky child survived. After one hundred and forty days at sea the ship put into the harbour of Honolulu. When it left a few days later three passengers were left behind in the island graveyard. Thomas Lowndes, miner, had died in the port as had one of the Ganner children. Mrs. Incher had given birth to a child in Honolulu and had died as a result. Her husband and the tiny infant sailed on towards Vancouver Island, but after a few days the motherless waif died. Mr. Gale, hardened seaman, wrote in his log the shocked and censurious note that "it was Throwen overboard, and no more Notice taken of it than if it had been a ded Cat." In mid-November the only child of

Richard Richardson was committed to the ocean not very far from the ship's destination. As though the sea had developed a taste for the dead of the *Princess Royal*, it refused to release the ship and tossed it for three days within the Strait of Juan de Fuca before allowing it to limp into the harbour of Esquimalt.[34] Chief Factor James Douglas reported to London with considerable satisfaction that in spite of a twelve-day delay in the Sandwich Islands the ship had taken only five and a half months for the journey. The passengers were in tolerable health except that two, a miner and a passenger, had been decently buried on the Sandwich Islands. Children often failed to survive in Victorian times, and he made no mention of the loss of seven children. He intended to send the miners off to Nanaimo immediately.[35] In a letter dated five days later he reported that they had already left for Colville Town in the ships *Beaver* and *Recovery*.[36] The name of the settlement was still giving Douglas trouble. The Spaniards had named the waterway on which the coal deposit stood "Boca de Winthuysen" and over time this had become corrupted to the English version, Wentuhuysen Inlet. The Indian name for the region was "Sne-ny-mo" and this the surveyor Pemberton adapted to Nanaimo which, Douglas told Barclay, would be used in future correspondence with London. But, in honour of the Company's current Governor, the settlement had already been dubbed Colville Town by the London headquarters. Such was the resulting confusion that Douglas, writing two letters in five days, used Nanaimo in one and Colville Town in the other. Not until 1861 was the confusion finally resolved and the accepted name became, and remained, Nanaimo.

What were the feelings of young John Thompson, miner, lately of Staffordshire, England, and his wife as they looked across the water to their new home? One can only guess. Relief, no doubt, that their long, uncomfortable and sometimes dangerous voyage was over at last. Wonder, maybe, that the settlement was so small and lost in what seemed to be a huge wilderness. Six months before they had been in the soot and grime and crowded clutter of the "Black Country"; filthy beyond belief but marked by a sense of historical continuity and permanence with everywhere obvious the mark of man's presence. Their new home was menacing in its impermanence and it seemed that the dark forest might advance at a whim and sweep the haphazard collection of buildings into the sea, leaving no trace. The only building of apparent strength and solidity was the building they had been told was the Bastion, a fort built for protection against the Indians. That purpose in itself implied danger. And yet the people waving on the shore did not seem troubled, in spite of the fact that only a short distance away stood a silent group of those same Indians. We can imagine the willing hands being held out to help them to land, friendly faces smiling a welcome to the newcomers from the outside world, friendly handclasps and an exchange of pleasant words. Old inhabitants and newcomers together must have slipped and stumbled up the wet track towards the houses, sharing the work of carrying the bundled belongings of the arrivals. Round about them children, free of the restrictions of a cramped ship, would have run and skipped and clambered over rocks in a delirious release of energy. And then there must have been an assignment of houses — and what houses they were. Mark Bate, who followed the Brierley miners to Nanaimo three years later, was to say:

> They told me how roughly the houses were constructed, of the dreary look outside, and the cramped space inside; how the chinks between the logs and poles, through which the wind would blow with a shriek of triumph, were plastered up with clay or stuffed with moss; of the interior equipage of benches, boards and bunk-like bedsteads; of the Dutch ovens for baking and cooking; of the drugget rush mats and rugs made in part of dog's hair by the Indians, used for floor covering.[37]

They were young and would survive discomfort. Just how young was shown by the census taken by James Douglas only a few weeks after they arrived.[38] Vancouver Island on the last day of 1854 had 744 white inhabitants; not many more than could be expected to live in a respectable English village. The tiny capital city of Victoria had 233 people living in seventy-nine houses, while in Nanaimo the forty-two houses contained 151 residents. It is difficult to think of a community in which only one person was over the age of thirty years, but that was the case in Nanaimo. Two thirds of the population of the mining village had not reached the age of twenty years and one third had not even reached ten years of age. Under twenty the balance of the sexes was almost even, but over twenty there were forty-eight men and only seven women. As on the day of arrival nineteen of the Brierley miners were married and there were already married men living in Nanaimo, it seems that, almost without exception, all the females in the "fifteen to under twenty" category of the census must have been married and, presumably, most of them already had children. There were twenty-nine children in attendance at the school where Mr. Baillie was the sole teacher. The school had been set up in the previous year when James Douglas had noticed that the children of the miners were "growing up in ignorance of their duties as Christians and as men."[39] He made no mention of girls and their duties as women.

If the sun really did shine on the Staffordshire party as they landed, it must have added to their hope for the future, and hope was about all they possessed. Given time, most of them would find a better life than the one they had left, but it would not be easy. The primitive mining village of Nanaimo was sadly lacking in sweetness and light; rather, its most obvious characteristics were dissatisfaction, trouble and turbulence. Perhaps the youthful nature of the community was part of the reason, but, whatever it was, Captain Stuart, as the Hudson's Bay officer-in-charge, was kept busy noting problems in his log.[40]

Nearing the end of the Brierley miners' first year of work, there was a strike in progress and Stuart noted that "York, Webb, Dunn, Harrison, Bull, John Baker and Incher —all in the strike — refusing to work." A day later the named miners had left Nanaimo, leaving their wives and children to fend for themselves. On the same day Captain Stuart heard that Meakin, loaded gun in hand, was searching for his wife so that he might shoot her. In an effort to avert a tragedy Stuart tried to handcuff Meakin. The authority and dignity of the Captain were somewhat bruised by some of the English miners who wrestled Meakin away, insisting that there should be a fair fight. Presumably the fight was to be between Meakin and another man rather than between Meakin and his wife, leaving the impression that even in a place so small, the opportunities for infidelity existed — or at least the suspicion of it if not the reality. No mention was made of the other interested participants, the results of the fight, or if it ever took place. Another day along and only a few miners were at work, "the rest being too lazy to work." It was time for action. At the end of a week, when the deserters had not returned, Captain Stuart had a simple message for the wives: the Company required their houses and they must leave them. A few days later all the deserters, with the exception of John Baker, returned from the coal mines of Bellingham Bay, or at least that is where Stuart presumed they had been. Such was the shortage of labour that they were all taken back as employees of the Company but, on the instructions of James Douglas, on terms less favourable than those they had previously enjoyed. It seems that the new terms did not sit well with Harrison, who was unmarried, and Incher, the widower whose child's body had been thrown overboard like a "ded Cat," for the very next day they left for California.

The pioneers were pioneering on. The first strike at Nanaimo had been called and lost. The miners had used their collective power, but it had not been effective. Those who had put up more than a token resistance to the Company and had chosen to try other places, only to return, had lost the most. The Company also had power to use; it had threatened the miners with eviction and had used its own economic advantage to improve its profits at their expense. Couple this with its use of force against the miners at Fort Rupert and the Hudson's Bay Company can be seen to have established a pattern of action and re-action between miners and owners that would continue until large scale coal production ceased on the coal coast.

There was no end to trouble, and, if trouble was in the offing, the truculent John McGregor would not be far away. He seemed to have had a genius for getting into scrapes from which he inevitably emerged the loser. Having one day been accused by George Robinson, mine manager, of neglect of duty, absenteeism and leaving work before time, he must have tried to get in the last word by calling after the manager "in most provocative and abusive language." Robinson, tired of words, ended the argument with action. As Stuart's journal records in a terse, descriptive, judgemental and medical entry, the end came when Robinson "picked up a hammer, hit McGregor on the head which knocked him down — provocation — wound contused rather than extensive." Swift action indicating that the Company representatives would stomach no foolishness.

Yet, in the face of threats, and, knowing what would happen if they were obliged to return to Nanaimo, miners still left looking for better conditions and broader opportunities. Some were lucky but most were not. The mining problems of the Hudson's Bay Company were nearly

George Robinson led the Brierley Hill miners as mine manager. Reluctant to suffer fools gladly, one argument ended swiftly when he "picked up a hammer, hit McGregor on the head, which knocked him down . . ."

at an end.

The grant of Vancouver Island to the care of the Hudson's Bay Company came to an end and was not renewed. James Douglas had to choose between the Company and the Colony and chose to be Governor of the Colony of Vancouver Island with additional responsibility for the new mainland colony of British Columbia. The clear separation of the affairs of government from the business dealings of the Hudson's Bay Company had been achieved and, although the Company retained its trading privileges, the main spring of its power on Vancouver Island was gone.

Coal had been the one bright spot in what had seemed a poor economic future for the colony, but the Hudson's Bay Company had made very poor use of its opportunities. Lack of knowledge of coal and of coal miners had played a part in its failure, but a contemporary writer suggested that the coal trade was so far from the normal investments of the Company that not enough attention was paid to it, with the result that there was no chance of a profitable operation.[41] In 1862 the coal interests were sold to the newly formed Vancouver Coal Mining and Land Company, an English concern, for forty thousand pounds. For this the new company received over six thousand acres of land, one hundred houses, and all the buildings associated with the mines. Within two years the output of coal doubled and at the first annual general meeting the chairman reported that the estimated coal reserves were three million tons. He went on to report that the first steam locomotive was on the island and at work helping to reduce loading costs; the shipments for the first six months of the current year were double those of the preceding six months, and so great was production that a lack of shipping space had made it necessary to charter extra ships to make continuous runs between Vancouver Island and San Francisco. Already the sale of one hundred land lots in Nanaimo gave cause to hope that the entire purchase price could be covered from the sale of land alone, as promised in the prospectus; in fact, all promises would have been met by a date two years after the date of ownership transfer from the Hudson's Bay Company.[42]

The continuing existence and growth of Nanaimo and the coal industry was assured. The pioneers had laid a secure foundation and they, and many others still to come, would build upon it.

4. *Robert Dunsmuir*

> Both the *Times* and the *Colonist* went so far as to suggest naming the island "Dunsmuir Island" and the latter paper urged that the Queen bestow a knighthood on the energetic industrialist.[1]

> . . . a man whose lust for wealth and power, and the opportunity afforded him to indulge in his semi-feudal aspirations and his hatred of common working people, would provoke sharp conflict in the mines, . . .[2]

The Robert Dunsmuir of popular myth is supposed to have begun his rise to fame and power, becoming British Columbia's most wealthy man, as a lowly miner at the coalface. It is doubtful that the myth matches the reality. His great-grandson, James Audain, claimed that when Dunsmuir was born in Scotland in 1825 it was to a family of "coal-masters."[3] A Scottish author of the nineteenth century in a book, whimsically but comprehensively, entitled *The Ayrshire Hermit: Tammie Raeburn, Hurlford Sixty Years Ago. Its Rise, Condition, and Celebrities, including Davis Clark the Centarian*, provides corroborating evidence by relating how Dunsmuir's grandfather, "a practical miner of good business habits," successfully worked a coal deposit at Skerrington. So well did the business prosper that in 1819 he opened a depot for the sale of coal in nearby Kilmarnock and a year later took over the lease of the Barleith coal deposits. His double lease having given him something of a local

Robert Dunsmuir: Pioneer immigrant to coal baron.

monopoly he celebrated by raising his prices. The comment that the elder Dunsmuir was "gradually becoming rich" no doubt accounted for his move from the small neat house at Old Hurlford to a house of two stories on the Riccarton Road.[4] However, the term "coal-master" was often applied to the aristocratic Scottish land-owners who had found coal on their extensive estates: the Dunsmuirs were certainly not of this class and the most likely comparison fitting their circumstances is that of the "charter-master" found in other parts of the British Isles.

Commonly a land-owner fortunate enough to find coal under his land would not concern himself with the details of its mining, but would contract with someone skilled in the task to raise a given quantity for a negotiated price.[5] The contract often went to a group of miners, working as equals, who appointed one of their number to be the main spokesman and negotiator. From there it was a short step to having a single miner negotiate the contract on his own behalf and, having been awarded the "charter," becoming the "charter-master" and deriving his profit from the difference between his mining expenses and the negotiated price. It followed that it was in the interest of the "charter-master" to keep wages and other costs to the lowest possible level. Clearly, he, who was once a miner, had become an employer, and to him would the complaints and hostility of the miners be directed about poor pay and conditions. It was between the contractor and the miner that the vital dividing line between capital and labour came to be drawn. The correspondence between "charter-master" and "coal-master" seems strong and, if this was the case, then it is certain that when young Robert Dunsmuir left Scotland for Vancouver Island, he was quite familiar with being on the side opposite to that of the miner at the coal-face.

The likelihood that Dunsmuir was not an ordinary miner is strengthened by the type and duration of his education. He attended Kilmarnock Academy until the age of sixteen, by which age the average boy of the village might well have been working underground in the mine for six or seven years. That Dunsmuir was so fortunate suggests that he was of a group more privileged than most. At sixteen he did enter mining as an apprentice to his uncle, Boyd Gilmour, but there is no record of the nature of his apprenticeship.

Many years after Dunsmuir had left Scotland, a letter sent to a newspaper during a bitter strike against him made a categorical denial that he had ever been a working miner: "As regards the extremely personal statement, regarding Mr. Dunsmuir having not many years since put in a daily shift as a working coal miner . . . it is untrue."[6] What evidence the writer had for making the statement and why he seemed so upset is unknown.

Dunsmuir has often been reviled for the tactics he used against his men when in later life he was a mine-owner on Vancouver Island, but it is unlikely that he was in any way being original. Within his own local community during the time of his Scottish apprenticeship he could hardly have failed to know of strikes and lockouts, of the importation of foreign blacklegs for strike breaking, of evictions and riots and the calling out of the Yeomanry to maintain the public order. He must have been aware of the rising power of the miners' unions, and the implacable opposition of the employers, which, in the case of nearby Ayr Colliery, took the form of a signed declaration that the employee renounced unions. To help them to avoid back-sliding, the agreement under which the men worked banned miners' meetings above or below ground and a man found, in spite of his declaration, to be a union member could expect to be severely fined by his employer. A grievance was a matter between man and management and no one else, and, if a man ceased to work for the company, his lot was immediate ejection from his company-owned house.[7] No union interference was tolerated. In 1850, the year Dunsmuir sailed for North America, a strike raged in his home village of Hurlford. The dispute was over a reduction in wages and the editor of the local newspaper wrote that, with trade depressed, it would be ruinous for the masters to pay higher wages and chided the miners for having "suffered themselves to be guided by those who make a trade of agitation."[8] The attitudes displayed and the resulting actions would all become depressingly familiar on Vancouver Island as the coal industry developed.

Robert Dunsmuir was a "transitional man" caught between times and places. He was firmly rooted in the traditions of a Scottish mining heritage, but blossomed into a rich and powerful man in the Victorian "Golden Age," which in itself was a time of transition such as never had been seen before. His background was in Europe, but his destiny was in a burgeoning America. That he attained his success away from Victorian Britain in no way lessens the importance of his roots or the influence of the British Isles in the development of the commercial pattern he was to follow so closely. Much has been made of the castle Dunsmuir is said to have promised his wife she would one day occupy as they left Scotland for Vancouver Island. Such a promise — if it was ever made — might well have illustrated his experiences in early life in Scotland more than it showed any hankering after grandeur and can be seen as a symbolic statement of his determination to some day be a mine owner. Mine owners, in his experience, commonly lived in castles. In Britain ownership of minerals had been left with the owner of the land and the landed aristocracy were naturally interested in the mineral producing potential of their estates. Having the greatest proportion of the land they obviously had the greatest chance of making coal finds and it was no accident that a great deal of the enterprise and capital of the coal mining industry came from the titled section of Scottish society.[9] Dunsmuir, in a sense, would be the inheritor of an aristocratic tradition and a dream of mine ownership could naturally lead to the assumption of a way of life which in its externals could be seen in the castles of his native land.

As Dunsmuir passed through his early years in Scotland there were undoubtedly important changes going on around him. The Industrial Revolution had brought obvious changes in the use of materials and in motive power: from wood, wind and water to the sure strength and power of steel and steam, but it also brought no less important changes in ways of doing business. Old ideas and attitudes were swept away and the commercial spirit became the principle of all action. Social sympathy was not compatible with that spirit; poverty and misery in the newly-created industrial cities were explained away by saying that the people were poor simply because they lacked the capacity and nature to be otherwise. Laissez-faire economics demanded that the industrialist or entrepreneur buy his factors of production as cheaply as possible and labour was included as one of these factors. The employer would employ any worker he chose, for as long as it suited his purpose and at the lowest possible wages — and think that he was doing the worker a favour by creating work for him to do at any price. The law of nature was that the strong should survive and prosper while the weak should perish, to the ultimate good of all. Social Darwinism, the survival of the economically fit, did away with the need for a social conscience and raised hard, ruthless selfishness to a high status. The poor had nothing to gain, but for the rich there was gain to be found in the changing balance of wealth and political power. The power of the aristocracy increasingly passed to the newly rich industrialists who owned not land but the factories and machines which were the new gold mines of the age. Men of great character, tenacious and unrelenting in their striving for the attainment of the single, uncomplicated end of commercial success came to dominate. To anyone not assured of rank by birth, such success was the way ahead and "to be a merchant prince was a far finer thing than being a gentleman."[10] The concept of hereditary status was shaken and it became a proud boast that "a respectable British citizen was on a level with any lord in the land."[11] Clearly, "respectable British citizen" was restricted in its application: while the poor, presumably not respectable, sank deeper into a quagmire of poverty, others, if they amassed sufficient wealth, were no longer born to their position but through their own efforts, and by seizing opportunity as it offered, could to a large extent achieve whatever they desired.

Robert Dunsmuir, looking back to the castles of the coal owners of his youth was, it seems, also absorbing the tenets of the commercial revolution which he would so rigorously apply on his rise to power and wealth. Often the ills of society have been made to resolve around personalities instead of around underlying social and economic forces, [12] and it is not to be an apologist for Robert Dunsmuir to say that he was a man of his time and no more to be judged by the morality of his modern critics than they are to be judged by the standards that will prevail near the end of the twenty-first century. He represents the archetypical entrepreneur of the Victorian Age and acted accordingly. Few men in his position in that era would have acted otherwise.

But Robert Dunsmuir was too late in his own country; the real opportunity for a man to start at the bottom and rise was gone. He would never have been able to amass sufficient money to break into the new aristocracy of wealth. His faculties would have to be used and his opportunities taken in a land still almost empty, where as yet there were no insurmountable barriers to commercial progress.

Joanna Olive White and Robert Dunsmuir were married in the Parish of Kilmarnock, County of Ayr, Scotland on October 1, 1847: oddly, the marriage certificate registered his name not as Dunsmuir but as "Dunsmore."[13] When the Dunsmuirs left Scotland for Vancouver Island three years later, they had two small daughters, Elizabeth and Agnes, and a third child was on the way.

A bald statement in a local newspaper carried the news that a party of miners was soon to leave Hurlford to search for coal on Vancouver's Island. It being likely that very few would know of the existence of the tiny colony, the further information was added that it was on the northwest coast of America, a six month journey away.[14] As the miners of the party were of no particular importance there was no mention of names. However, later a report described the presentation of a "valuable double barrelled gun and a revolving pistol" to Boyd Gilmour, Dunsmuir's uncle, who was to lead the party. The morning after the presentation the party left on the first train out of Hurlford for London, struggling to maintain a calm exterior but betraying their emotion by tremulous accents when they spoke. The report concluded "On Friday they embarked at Gravesend, on a voyage of not less than 15,000 miles, to regions where half the convex globe will intervene between them and their native land." [15]

The voyage was not without its tribulations. After safely rounding the perilous Cape Horn the *Pekin* came ignominiously to rest on the

Joan Dunsmuir: promised a castle on leaving Scotland.

sandbar at the mouth of the Columbia River. Fort Vancouver then became an intermediate stopping place. It was there that the Dunsmuir's first son, James, was born in 1851. The journey was eventually resumed and they were carried on to Fort Rupert in the ship *Mary Dare*, arriving there almost a full year after leaving Scotland. These successors of the Muirs had no better luck in finding coal deposits of value and soon were sent down the coast to the better prospects of the new settlement of Nanaimo. After a short while the contract of Gilmour came to an end and he returned to Scotland. The Dunsmuirs decided to stay; a decision fraught with implications, both good and bad, for many generations of coal miners and their families.

Again the question of whether or not Dunsmuir was a miner among miners arises, for he had not been long in Nanaimo when James Douglas mentioned in a letter that "now Dunsmuir has a small percentage upon the extra coal produced"[16],which suggests some kind of supervisory position. But it was during the turbulent first year of the Brierley miners presence that Dunsmuir made the first significant step towards independence. While the rest of the miners were striking, he steadfastly refused to join them and quietly made arrangements through which he would be able to work without interference. On Friday, October 12 of the year 1855, Captain Stuart logged in the official journal of the fort the comment that "Dunsmuir commenced working on his own account"[17] and, by the time James Douglas had arrived to sign the formal agreement, Dunsmuir had already shipped his first cargo of coal to San Francisco on the *Otter*.

In the next few years Robert Dunsmuir did not become rich. Few in Nanaimo did, it seems, for in the elections of 1859 only one person met the voting requirement of owning twenty acres of land.[18] In 1862 the Hudson's Bay Company sold out to the newly formed Vancouver Mining and Land Company, bringing to an end Dunsmuir's agreement to work "on his own account" and he joined the new company as mine supervisor. Two years later he became resident manager of another new company, the Harewood Coal Company, a venture which was to have only a brief life. Wisely, Dunsmuir left the company before the end came and went back to his old job of mine supervisor. At that point in his life there was a marked contrast between how the miners felt about him and how they would later come to regard him, for, on leaving the Harewood Company, the men he had supervised bought him a gold watch. The testimonial dinner, at which the watch was presented, ended with three cheers and the singing of "For He's a Jolly Good Fellow."[19] No doubt some of the men present that night would later recall the event with a sense of wonder.

By that time life must have been relatively easy for the Dunsmuirs, in spite of a large and growing family, but still Robert Dunsmuir added to his income by growing vegetables for sale to visiting warships and by dabbling in real estate.[20] A March, 1866 edition of the *Nanaimo Free Press* contained the notice:

> FOR SALE: Those desiring building lots with frontage on Commercial and Bastion Streets, being Lots 2 and 3 of Block 59 the most elegible building lots in Nanaimo (for terms which are extremely moderate), apply to Robert Dunsmuir.[21]

When Robert Dunsmuir made his next important advance he was in his mid-forties and had been on Vancouver Island for eighteen years. His progress towards living in a castle had not been very spectacular.

According to his son, James, Robert Dunsmuir never forgot his dream of finding a coal seam of his own and seldom stopped searching. And then, one day, he returned from a fishing trip and told his wife that he thought he had found coal.[22] In his own version of the discovery of the Wellington Seam that was to make him rich Robert Dunsmuir wrote:

> When I was in the bush, in the month of October, 1869 . . . I came across a ridge of rock which I knew to be the strata overlying the lowest seam that has yet been discovered here. A short time afterward I sent two men to prospect, and in three days discovered a seam of coal 3 1/2 feet in thickness, 30 feet below the tops of the ridge, dipping S.E., one foot in six.
>
> After procuring from the government a right to further prospect, I sank a slope . . . and mined therefrom about five hundred tons, twenty-five tons of which were taken onboard of *HMS Boxer* for trial. The same quantities were taken from the Vancouver Coal Company's Douglas Pit and the Newcastle mine.[23]

The report of the sea trials of the coal favoured the Dunsmuir coal; the foundations of the castle had been laid. But discovering a seam of coal, however rich, is not quite the same as having a producing coal mine. The coal lay thick and extensive a few miles to the northwest of Nanaimo. And there the bulk of it would stay unless Robert Dunsmuir could tap a seam of money far beyond what he could hope to raise personally. After an initial and rather discouraging attempt to get sufficient capital, his final solution came from the British Navy, which seemed to be replete with officers having considerable sums of money at their disposal.

On November 24, 1871 Robert Dunsmuir, Lieutenant Diggle of *HMS Boxer* and Admiral Farquhar met, in the presence of Magistrate Spalding, and signed a partnership agreement which stated that for a period of twenty-five years they would be "partners in the trade and business of Coalminers and coal proprietors and Merchants." Between them they were to supply, in cash or kind, the equivalent of thirty-two thousand dollars; ten thousand to come from Dunsmuir, ten thousand from Wadham Neston Diggle and twelve thousand from Arthur Farquhar. For these sums they were to be "interested in equal shares" with the provisos that for acting as mine manager Dunsmuir would get an extra ten cents for each ton raised and Diggle would get the same for each ton of coal he sold as the Company's Victoria agent.[24] It was not stated in the agreement but it was understood, as it would be in all Dunsmuir's later ventures, that he had complete control of the operation. An immediate need was to have a clear and unrestricted access to the shore of Departure Bay and one of the first acts of the new partners was to rent a strip of land thirty-three feet wide from the edge of the mine property to the shore. At the seaward end of the strip an acre of land was leased as the coal shipment point. That not everyone was interested in coal was obvious from the terms of the agreement which stipulated that the strip should be left unfenced so that cattle could freely pass. The owner also insisted that intoxicating liquor should not be sold on the land.[25]

With growing prosperity it was time for Dunsmuir to provide his family with better accommodation, and so the house *Ardoon* was built. There was no question that space was needed, for eventually he would have to house Elizabeth and Agnes, the daughters who made the long journey from Scotland as very small girls; son James born on the outward voyage after their rescue from the *Pekin* aground on the bar of the Columbia River; Alexander, the first white child born in Nanaimo; and, in order of birth, daughters Marion, May, Emily, Jessie, Effie and Maude.

The first shipment of coal from the Wellington Mine was carried to San Francisco on the *Caesarwich* and there reached the consumer at a price of $22 per ton.[26] The gap between what Dunsmuir was paid for the coal and the consumer price was so great that Robert Dunsmuir immediately decided that one day he would set up his own San Francisco office so that he might have a better share of the ultimate price. This he did some years later with his son Alexander as his representative.

By 1873 the Wellington Colliery was producing a coal tonnage one third as great as that of the rival, and long estabished, Vancouver Coal Mining and Land Company. The Nanaimo mines were booming: in 1876 the total production of the area was 76,000 tons. Two years later the Wellington Mine alone produced a black flood of 88,000 tons — several thousand tons ahead of the production of its rival. From every ton Dunsmuir was becoming richer. Within ten years of the discovery of the Wellington Seam Dunsmuir's company had four locomotives hauling the massive output over almost five miles of track to the three company loading wharves. Then with the purchase of the Chandler Mine to the south of Nanaimo the length of track doubled, another locomotive was added to the stable, and a total of 450 coal wagons were transporting coal to the dockside.[27] In 1881 the Wellington Colliery produced almost 190,000 tons and at the end of the year its slim stockpile had grown by less than half a day's production, indicating that almost every scrap of coal mined throughout the year had been sold. This vast quantity of coal was dug and processed by the combined efforts of 259 white workers and 266 who were Chinese — the balance had tipped in favour of the Chinese and the reason was not difficult to find. Miners at the coalface were being paid daily wages between $3 and $4; other white employees were earning between $2 and $3.75, while the Chinese were getting between $1 and $1.25.[28] Dunsmuir was following the accepted practice of buying in the cheapest labour market with a vengeance. The value of the company's plant, at the time when the Chinese were being paid wages on which a white worker would have starved, was around a quarter of a million dollars, a vast sum for that age. Dunsmuir was rich and would be richer yet, for the profits by this time had to be shared among fewer people. Admiral Farquhar and Captain Egerton who had invested capital in Dunsmuir, Diggle and Co., had been bought out and only Diggle remained as an outsider to the Dunsmuir family. It must have been with a great deal of satisfaction that Robert Dunsmuir, at the end of 1883, announced in the *Colonist* that:

> Mr.N.W.Diggle, having withdrawn from the firm of Dunsmuir, Diggle and Co., doing business at Departure Bay, Nanaimo, the business in future (will) be carried on under the name and style of R. Dunsmuir and Sons.[29]

Robert Dunsmuir was now truly independent, but the price was high. Diggle agreed to the absolute sale of all his interests in the partnership for $600,000 of which sum $100,000 was to be paid immediately with the rest to come within five years, and during the payment period Diggle was to receive 5% interest on the remaining balance.[30] Seldom can any man have become so rich on an initial investment of $10,000 — unless it was Dunsmuir himself. It was remarkable that Dunsmuir was even able to contemplate the payment

Wadham Neston Diggle of Dunsmuir, Diggle & Co. — in for $10,000, out for $600,000.

of such a sum of money in an industry of which it had been said only a few years earlier in response to a threatened strike for higher wages that " . . . every coalmine on the coast is losing money."[31]

Robert Dunsmuir's greatest triumph was yet to come. It came by a business stroke seemingly unconnected with coal, which nevertheless transformed his coal holdings into an empire.

Canada was to be bound together with rails of steel. British Columbia, child of the railroad, was to be tied in within a guaranteed ten years and with a start on construction to be made not later than two years after the Confederation date of July, 1871. As the beginning date approached Sir John A. Macdonald, having nothing better to offer, grandly announced that the western terminus of the Canadian Pacific Railway would be Esquimalt, the naval base town adjoining Victoria. On the day before the two year deadline was to pass, with appropriate and dignified ceremony, a group of local worthies turned the first sod at Esquimalt — if that constituted a beginning then the Prime Minister of Canada had kept his word. The choice of terminus carried with it an automatic assumption of the position of the mainland end of the railroad: it would have to be Bute Inlet. The

Coal mine in the forest — Dunsmuir's No. 1 Slope, 1875.

tracks would cling to the precipitous slopes of the fiord until it was possible to advance no further and then they would leap-frog from island to island on bridges built across the intervening tide races until the shore of Vancouver Island was reached and there the tracks would swing down the eastern shore through Nanaimo and on to Esquimalt. It was a route strewn with engineering problems; impractical, but with powerful advocates. At the end of the year in which the first sod was turned at Esquimalt there were, in fact, six different routes to the coast still being examined. Two years later the number had grown to twelve. Even within British Columbia there was no agreement as to where the rails should end: a vociferous and determined group on the mainland were as adamant that the Fraser River - Burrard Inlet route should be followed as the inhabitants of the island were determined it should not.[32] A railway on Vancouver Island, in the opinion of the editor of the *Mainland Guardian*, was "about as necessary as a fifth wheel on a cart."[33]

For years the railway disputes dragged on, through explorations, engineering studies and cost comparisons. They continued through clouds of suspicion and mistrust, through offers of reparations, attempts at arbitration and thunderous threats of secession, until at last the Federal Government decided on the Fraser Valley-Burrard Inlet route. Vancouver Island would not be the western terminus, but by the time the decision had been made it already had a railroad. Something had to be seen to be done during the long years of indecision; political debts to the capital city had to be repaid. The seventy-eight miles of railroad between Esquimalt and Nanaimo was the price. This was the age of the railway mania when any length of line was seen as a guarantee of development and progress; an age in which politicians were "willing to trade the kingdom for a horse, an iron horse."[34] They were ready to give away the island kingdom as a subsidy for building a short length of line, but first a recipient had to be found.

Robert Dunsmuir, the island's richest and most powerful businessman, was an obvious choice to build the line. He was not, it is true, a builder of railways, but he had experience in business and no one doubted his organizational capacity and driving energy. At first sight there seemed to be little reason for him to be interested, as a line from Nanaimo to Esquimalt would be of no advantage to his coal business. He had his own short lines, which were quite capable of moving the product of his mines the few miles from the pit-head to his wharves, where the coal was loaded directly onto ships for export. But he was a businessman and there was gain in the offing, and great gain when it became clear that he could almost name his own price for building the line. He set a steep — even breath-taking — price. He wanted $750,000 in cash to start and then in addition would have to be given 1,900,000 acres of land which somehow within its boundaries just happened to contain all the known coal reserves of the island. Not only was this tremendous area of land demanded but also everything connected with it: "minerals . . . in, on or under the lands," all rights to the foreshore, the right to mine under adjacent sea beds and the right to cut forests at will. He would want, and get, the right to erect any building he desired and freedom from taxation during the construction of the railroad.[35] If Dunsmuir was to be given the contract, not even his riches would be sufficient to finance the entire project and partners were needed. So it was that when the agreement was signed the names of the sub-contractors were Charles Crocker, Mark Hopkins, and Collis Porter Huntington; Americans of the South Pacific Railroad who were not known for their ethical business practices. The contractors named were Robert Dunsmuir, his son, James, and his son-in-law, John Bryden.

Vancouver Coal Mining and Land Company's Pithead No. 1, Nanaimo.

Against considerable parliamentary opposition, Sir John A. Macdonald steered the "give-away" through the federal parliament. Member of Parliament D.W.Gordon of Vancouver wished to know if tenders had been requested through government advertising for construction of the railroad. Sir John replied that they had not. To the charge that a small clique was being handed the monopoly of the coal deposits on the island, it was answered that there was a precedent. (The precedent turned out to be a single square mile of Nova Scotia). The involvement of American citizens in this gift of a part of Canada was not looked on with any great favour by many, but, as the Dunsmuirs were said to own three-fifths of the company, it was claimed to be under the control of Canadian rather than American capitalists. Dunsmuir had done it again: control was in his hands — he had power greater than even he had ever dreamed possible. He had the island in the palm of his hand. In the land grant area only the land and mineral rights established under previous crown grants escaped his net. And even then it is said that he tried to claim the coal rights on land that was not fully paid for at the transfer date.[36]

Robert Dunsmuir was not a man to waste words; he made his meaning clear and expected action. The only instruction he ever appears to have given about the construction of the railroad was contained in a letter to Joseph Hunter, Engineer-in-Chief, which read:

> As Engineer in Chief you will at once proceed with the location of the line between Esquimalt and Nanaimo in accordance with the agreement entered into between the Dominion Government and the Esquimalt and Nanaimo Railway Company of which Company I am President.
>
> You will be expected to locate the line with the view to the least expense in construction consistent with good work, and carry out the work to completion without delay. You will be responsible for all expenditures in your department returning proper vouchers duly signed to this office, and report to me your progress in the work as often as possible.[37]

Hunter had his instructions, and there was nothing more to be said. Now Dunsmuir wanted action and he got it. In August, 1886 Sir John A. Macdonald drove in a "last spike" to mark the official completion — ahead of schedule.

Dunsmuir was also a politician, though one of no great distinction. In 1882 he entered the Provincial Legislature as Nanaimo's representative. He was re-elected in 1886, at which time he became President of the Executive Council. It was during his second term that two members tried to call him to account for his apparent "disloyalty" after he stated in a speech made in Portland, Oregon that it would be to his financial interest if Vancouver Island was to become a part of the United States. That statement could have been taken simply as a matter of fact; however, Dunsmuir went on to say that, apart from his personal situation, it would in any case be better for British Columbia because its commercial interests were with the United States. His business interests no doubt were wider than national boundaries; all countries to him had the common denominator of money. The enquiry got nowhere and Dunsmuir was cleared.[38]

Robert Dunsmuir ruled an empire but still lacked a castle. On top of a small hill in the midst of a twenty-seven acre estate in Victoria he began to build a baronial mansion of stone; solid, turreted, with tall chimneys reaching above the steep red roofs. Such a mansion as would not be out of place in the midst of Scottish heather moors. The interior was rich in solid wood panelling — oak, mahogany, teak and walnut. The furniture was heavy, hand-crafted and seemingly designed to last forever. Every room had a fireplace, and some had two, elegantly wrought and with no two alike — there would be no shortage of coal to fuel them in the Dunsmuir household. Stained glass was there in abundance, in the fifty-foot living room casting a multi-hued light on the crystal of the chandeliers. There was a billiard room and a dancing room upstairs reached by a massive oaken staircase, intricately and elaborately carved. The staircase ended at a tiny room at the top of a tower from the windows of which could be seen the Strait of Juan de Fuca backed by the Olympic Mountains, while in other directions, framed by steeply pitched roofs and tall, thin chimneys, could be seen panoramas of the interior of the island. No expense was spared to build Craigdarroch Castle, named after Annie Laurie's Craigdarroch House in the Scottish native land of the Dunsmuirs. Others have since seen the castle not as Robert Dunsmuir saw it but as a monument: "a monstrous headstone standing over the graves of miners who died at the coal-face, or at home in agony, from the effects of 'miner's consumption'; a crude monument to their wives and children who went hungry to the grave."[39]

In 1890 Joan Dunsmuir moved into the castle her husband was said to have promised her as they left Scotland. It had taken forty years for the promise to be kept. In that same year she donated three stained glass windows to St Andrew's Presbyterian Church in Victoria in memory of her husband for she moved into Craigdarroch Castle as a widow.

A severe cold confined Robert Dunsmuir to his bed in April, 1889. Early one morning he was found in a "death-like state" from which he was revived only to have a sudden fatal relapse a few days later. At the time of his death he was 64 years old, a pioneer who had lived on Vancouver Island for almost forty years. His rise to wealth had been meteoric: most of what he had achieved and acquired had come to him in the last third of his life. Arriving as a lowly employee of the Hudson's Bay Company, he left in death as the island's largest owner of coalmines, the owner of a fleet of steamers and sailing ships, most of the Albion Iron Works was his, he was President of the Esquimalt and Nanaimo Railway and, among other things, he was chief shareholder in the Victoria Theatre.[40] In the announcement of Dunsmuir's death the editor of the *Colonist* expressed the opinion that he had "scarcely an enemy in the world,"[41] an opinion which spoke no ill of the dead but one which was highly questionable. Dunsmuir's actions had always provoked controversy; by some he was revered and by others hated, and the difference mainly depended on the relationship of the speaker to Dunsmuir. He was far from being universally loved, but it is possible that he was universally respected.

On the day of the funeral, while Robert Dunsmuir lay under the plate glass lid of a coffin of metal embellished by fittings of massive silver, thousands of people were taking up their places along the funeral route. Over a thousand passengers had come by train from places along the Esquimalt and Nanaimo Railway using free passes edged in black and bearing the words "In Memoriam" on the reverse side. The funeral procession contained twelve hundred men. Of the family, James and Alexander Dunsmuir rode in the carriage immediately after the horse-drawn hearse. Behind them came the four Dunsmuir sons-in-law. As the remains of Robert Dunsmuir were lowered into the grave and earth was cast on the coffin, the Wellington Colliery Band played "Nearer My God to Thee," the music swelling and echoing across the waters of the bay. In the procession the miners of Wellington were the nearest of all the Dunsmuir workers to the body of the man whom they had so often fought and railed against.[42]

Only a few months before his death some of the men from the Wellington Mine had visited Robert Dunsmuir as a deputation to ask why he had closed the mine. He told them the closure was because there was a threat of a strike. Strikers were never taken back into his mines. By closing the mine he had prevented the strike and so could take back the men in whom he had confidence when the mine re-opened. He went on to say, "I did not want it said that I had banished little children and women in cold winter days without giving them who support them a chance, and so I shut down the works for that reason and nothing else . . . I consider I was thinking more of your interests than you yourselves." Perhaps he believed it. Characteristically, Dunsmuir ended by saying, "Tell those men who would strike that I am a stubborn Scotsman and that a multitude cannot coerce or drive me."[43]

In a will of startling simplicity for one so rich Robert Dunsmuir stated: "I give, devise and bequeath all my property both real and

Main "watering hole" of short-lived Wellington.

personal to my dear wife Joan Olive Dunsmuir absolutely and appoint her sole executrix of this my will."[44] His legacy to others, though unwritten, was much more extensive: from the time of his discovery of the Wellington Seam he was the dominant figure in Vancouver Island coal and the patterns of employee relationships he had established would haunt the industry until its end.

5. Strike

The Coal Miner

And then we cursed ill air and reek,
I'm keepit gaspin' a' the week;
For which my recompense is small,
In fact it's next to nought ava.

For wages more I've tried the gaffer,
Ye might as well speak to a Kaffer;
'Begone and be content,' says he,
'two cents mair I winna gie.'

But when my pay comes round, and me
Gets't in my hand I'll let him see;
He'll no oppress and taunt me baith
I'll close the place and draw my graith.[1]

There was an inevitability in the way the miners helped to create their own future scourge. In striking against the Hudson's Bay Company at Fort Rupert the Muirs made it necessary to have replacements sent in and one of them was Robert Dunsmuir. The Brierley Hill miners, striking in their dissatisfaction with conditions at Nanaimo, made decisive the distinction between Dunsmuir and miners generally, for it was in this period that he, not striking and having no patience with what was going on, received permission from James Douglas to "work on his own account."[2] The story of the coal industry on Vancouver Island is one of strikes; strikes bitter, violent and often protracted. Strikes were called in every mine and against every employer, but most often, and in their most extreme form, they were fought against Robert Dunsmuir.

The early miners' strikes lacked direction and organization. There were no unions and while the simple expedient of desertion was possible there would be none. Desertion had been the way of relieving personal dissatisfaction in the fur trading days; it had been used when the Muirs left Fort Rupert; it had continued in the early days of Nanaimo, and beyond. There were always attractive lures: the goldfields of California or the gold rush on the mainland only a few miles across the water; always something to relieve the need for concerted action. But the far green fields were often disappointing, men returned with discouraging stories. The mining population grew, the outside attractions lessened, the need to stay and put right what was wrong increased. In the main, the miners were immigrants from the British Isles brought up in a society of rigid divisions — the owners of capital and the providers of labour. They had in their own past experience struggled and organized unions to improve their conditions of work, knowing well that united action was their only hope. The Vancouver Island situation had come to be not so very different from what they had known in the British Isles — the capitalists were becoming rich at the miners' expense. All that was required for union action was a trigger and that Robert Dunsmuir would provide.

The strike of the Wellington miners was on. Or was it? The miners had no doubt of the fact, but Robert Dunsmuir was adamant that no strike existed. As far as he was concerned, a strike threat had caused him to dismiss the men and close the mine. They no longer worked for him and so there could be no question of a strike. The men had no need to think that they would get their jobs back again or that he would give in to their demands. The men he had dismissed were finished forever and his notice in the *Nanaimo Free Press*, repeated for emphasis, made that abundantly clear:

> There is an impression in the community that we are obliged to accede to the miners' demands, but for the benefit of those whom it may concern, we wish to state

> publicly, that we have no intention to ask any of them to work for us again at any price.[3]

Strike or not, there was no work at the mine, and on Monday, February 26, 1877 the miners of all the collieries of the Nanaimo area met to hear what the miners of the Wellington Colliery had to say.[4] Lacking a formal miners' organization, there were uncertainties and fumblings until at last a Wellington miner, MacDonald, was voted into the chair with McKenzie, a Harewood man, appointed to act as secretary.

As MacDonald rose to his feet, the hubbub gradually stilled and he opened the proceedings. "Gentlemen, you all know we are on strike and you may have heard why: twenty cents a ton was taken from us, the weight we are given is short and we must do jobs about the mine which should not be done by skilled miners. Foreign labour was brought in but, thankfully, was persuaded to leave peacefully. I am glad to say that the men stood firm for their rights but kept within the law and so have kept the sympathy of the community. I call upon George Vipont to speak and tell what he knows of the situation at the Wellington Colliery."

George Vipont rose and looked around the smoky hall before beginning to speak. He was a comparative newcomer, having moved from the United States only three years before. Conditions had been good when he arrived, but then, as always seemed to happen, the mine owner had become greedy and things started to get worse. He had a wife and family to think of, as had most of the other men, and, if he could not make a living wage, they would suffer along with him. He had supported the strike when it was first suggested earlier in the month. After what Dunsmuir had done since, he was now firmer than ever in his support. He spoke briefly, directly and without evasion. "Mr. Dunsmuir is an untruthful man. He reduced our pay for each ton of coal mined from the old rate of $1.20 to $1 saying he was sorry to do it and would go back to the old rate as soon as the market improved without us having to ask him. The market in San Francisco has now improved, but he still refuses to pay more than $1 per ton. When the pay went down he promised the price for blasting powder would go down 50 cents a keg — it never did. Even then we would not have gone on strike had it not been for the short weight. The cars increased in size while the weight stayed the same. No matter how high we piled the coal the weight never showed any higher on the scale.

Since the strike began Dunsmuir has built a gate to keep out the grocer who charges lower prices than the Wellington store; the baker has been warned to bake us no bread and the butcher to supply us with no beef. When he cautions the doctor not to visit the sick for he surely shall not be paid for his services, he attacks our wives and children who have done him no harm. He tries to starve us out and leaves us to rot while he drives by in his carriage. Dunsmuir has brought in strike-breakers, the sweepings of the San Francisco gutters, and it has cost us money we can ill-afford to get rid of them. We are fighting for our livelihood and for our rights as men; I say we must organize a union and act as one."

As Vipont sat down again there was thunderous applause from the Wellington miners and a good deal of support from miners of the other pits. They were all familiar with the kind of treatment of which Vipont complained and could sympathize. But some men held back — a union? Some men were unsure of unions and saw them as possible limitations on their own freedom as well as on the actions of the bosses. Mr. Clarke spoke next. "This kind of trouble is not new and has caused many strikes. In Illinois when we made complaints about the weight we were told that if we didn't like it we could quit and get out of the company houses. We put a stop to the disputes by appointing a check-weighman to look after the miners' interests. That could be done here also." Some men nodded agreement; others looked doubtful, thinking that to persuade Dunsmuir to agree to such an arrangement would be difficult, if not impossible.

From the waving mass of arms below him, MacDonald selected one of the most vigorous and identified its owner. "Mr. Hockyard has the floor." Hockyard was not a Wellington miner but he was determined that they must have their rights; fairly if possible but if that could not be, then by any means. He said, "Dunsmuir has cheated these men in every way he could, like all bosses. We get nothing from the bosses that they can keep back from us. If we hope to get decent conditions we have to beat them and that we can do only by banding together and forming a union."

Discussion continued. Most men were in favour of forming a union, but some still had doubts. All were definitely on the side of the Wellington miners and against Dunsmuir, but was a union the way to solve the problem? Arguments swayed backwards and forwards; for and against, until the discussion became repetitive, circular, pointless. MacDonald at last tried to resolve the issue by saying, "Our great want is for a society for the protection of labour, led by men with cool heads. Will someone make a motion?" A resolution was proposed "That the miners of Vancouver Island do form themselves into a Mutual Protective Society, so that labour and capital will work harmoniously." The motion passed by a large majority and before they left the hall, 137 miners signed up as members.

The suggestion that the suppliers of labour and the owners of capital could work together harmoniously was no doubt an attempt to word the motion in such a way as to avoid provocation rather than an expression of the miners' true beliefs. They were not so naive. Replying to the question, "Is it not in the interest of employer and employee to work together?" the secretary of the Amalgamated Society of Engineers, appearing before a British Royal Commission ten years earlier, had said, "There I beg to differ. Every day of the week I hear that interests are identical. I scarcely see how that can be in a society which recognizes the principle of buying in the cheapest and selling in the dearest market. It is in their interest to get labour done at as low a rate as they possibly can, and it is ours to get as high a rate of wages as possible and you can never reconcile those two things."[5] For the age a very realistic view grounded in bitter experience and a view Robert Dunsmuir would have understood well.

The miners would find that there was a vast difference between having a union to carry out negotiations and getting employers to negotiate. Twenty-five years after the 1877 Strike and the formation of the union, Francis Deans Little, the general manager of the Wellington Colliery would still be disclaiming knowledge of whether or not his coalminers were union members in a manner suggesting that he didn't care very much either.[6] In this he was simply following the tradition of Robert Dunsmuir who handled unions very simply and effectively by ignoring their existence. He would accept no outside interference, no negotiators, no arbitrators: he was the master of his own business by his own efforts, the Grace of God, and economic might, and there was an end to it. He would talk to his own men, which at least put him ahead of many Victorian employers who considered it humiliating to have to meet workmen round a common table, but not on terms of a non-existent equality. He would meet them at his convenience, in a place of his choosing and with the odds very firmly stacked on his side. But, very definitely, he would have nothing to do with unions.

There was no doubt that the miners had stung Dunsmuir by forming a union. He was ready to set about those who had particularly displeased him. Vipont, with so much to say at the miners' meeting, became a special target. A lengthy "Card from Mr. Dunsmuir" appeared in both the *Nanaimo Free Press*[7] and the *Daily British Colonist*[8], repeated so that no one should miss it, in which George Vipont and "his tissue of falsehoods" were systematically dissected. The claim made by Vipont that nothing could get the scale to read over ten hundredweight was clearly false: the average weight of his boxes was listed from July 1876 to February, 1877 and for each month the average weight was over ten hundredweight. A further table was provided that showed the average weights of an unidentified miner and all the weights were above Vipont's averages: Mr. Dunsmuir icily suggested that if Vipont had filled his boxes similarly the weighing machine would have given him higher weight also. As to having built a gate (to keep out the grocer) on his private property, "Is it any business of his suppose we build a stone wall eighteen feet high around every inch of it?" When Vipont broke his leg — through his own laziness and carelessness, the mine owner pointed out — Dunsmuir claimed to have helped him until he was able to work again. When allowed to contract for work Vipont was soon unable to pay his workers and he, Robert Dunsmuir, had set aside Vipont's contract and paid the men what they were owed. He could not understand why Vipont should hold such enmity towards him, unless it was because he had loaned him money to bring his family from the United States. The money was paid back over a long period and in small sums, but "if you lend anyone money at any time he is sure to be your enemy." A perfect piece of character assassination had been carried out: George Vipont had been made to seem a liar, he was quite obviously both lazy and careless, incompetent when it came to running a simple venture, was ready enough to borrow money, but paid back slowly. Vipont was vindictive and untrustworthy and the implication was that if such a man supported the union then it must be likewise. Robert Dunsmuir took the opportunity to express his bewilderment that the miners had ever decided to strike in the first place. He knew of no grievance about the twenty cents per ton and the weighing machine had been adjusted to the satisfaction of the men. There might have been a cause for grievance in that seven boxes of coal were to be considered a day's work, but many men had said they could produce eight boxes and one man claimed he could send out ten boxes daily. There was, it seemed to him, very little substance behind the miners' complaints.

Dunsmuir was not finished with Vipont yet. Along with Uriah Hockyard, who had the temerity to accuse Dunsmuir of cheating his men, he was hauled before the local magistrate "for having procured the publication of his speech in the *Nanaimo Free Press*."[9] In court, Mr. Barnston, for the defence, raised an immediate objection on the grounds that the presiding magistrate, Captain Spalding, was the legal representative of Admiral Farquhar, a partner in the firm of Dunsmuir, Diggle and Co. This accusation was indignantly denied by the Captain, and the case, weak though it was, proceeded. It was easily shown that it was the desire of the entire miners' meeting that the press should cover the event and to that end an adjournment had been called to give the reporter time to get to the meeting. The charge laid against Hockyard was dismissed and that against Vipont dropped. Then the prosecution and the defence changed roles and Barnston, prosecuting on behalf of John Bolton, charged Dunsmuir, Diggle and Co. with "using false and defective weights."[10] For the defence Mr. Davie argued that the Dominion Statute regarding weights and measures was not in force in the province. On the plea that he was unsure and would have to consult the Attorney General, Captain Spalding postponed the hearing. From that point the matter seems to have been forgotten. So far in the game there had been no score. But the charges stemming from the meeting were not yet over — George Norris, proprietor and publisher of the *Nanaimo Free Press*, had to answer a charge of libel preferred by Robert Dunsmuir. The libel case delved deeply into the accuracy of the charges made at the miners' meeting, which Norris had reported in his newspaper, and much attention was given to the matter of the allegedly incorrect weights. Through a wearying length of question and answer, charge and counter-charge, corroboration and contradiction, the case ploughed on until Norris was eventually committed for trial. Immediately Dunsmuir dropped his complaint of libel on the grounds that he wished only to clear the characters of the members of his company and felt that the evidence given at the preliminary hearing had adequately disproved all the charges made against Dunsmuir, Diggle & Co.[11]

The legal actions, however, were far from over. Chairman MacDonald had stated at the meeting, with some pride, that the strike-breakers who had been brought in by Dunsmuir had been persuaded to leave by law-abiding persons "with no law officers after them."[12] He was a little premature. It should have been so, but it was not. When twenty-six of the prospective strike-breakers arrived back in Victoria on their way home to San Francisco, they said that they had declined to work in the mine at Wellington after hearing of the true state of affairs. In spite of this, one of their number swore out a complaint. As a result four miners from Wellington — Henry Adams, Henry Smith, John Jenkins and James Knight — were brought before Judge Courtney in the Victoria court charged with intimidation. The story, as related in the evidence of James Knight, was that at a public meeting the San Franciscans had been asked to divide themselves up according to those who wanted to return home and those who did not. Twenty-six chose to leave and had requested an escort to Victoria as they were unfamiliar with the coast. The four accused had been chosen as escort. The miners at Wellington had paid the board of the men while they were in the Nanaimo area, had paid the rental of the canoes in which they had travelled to Victoria, and had money to pay the fare of each man back home to California. Knight knew of no threats being made. His evidence was confirmed by one of the erstwhile strike-breakers who thought that the accuser, Osborn, had been in no danger at anytime. In passing he remarked that, "Dunsmuir sent into the saloon for Osborn," leaving a distinct impression that the charge had been laid at Dunsmuir's instigation.

In reply to questioning by the striker's lawyer, Osborn clung to the story that he was a victim of force and that his life had been threatened. He admitted that he had made no complaint in the canoe, even though he was one of a party of twenty-six opposed to four miners, explaining he was in fear of his life and was not sure that his comrades would have protected him in case of trouble. Mr. Justice Courtney released the accused without a blemish on their reputations, after first apologizing for having confined them. He went on to make a remark about Dunsmuir and Co. being honourable men, phrased in such an ambiguous way that it could well have been taken to mean that he suspected they had a finger in the pie somewhere. It seemed that the reputation and influence of Robert Dunsmuir was not as great as it was destined to become in Victoria, for the spectators in court greeted with applause the judge's decision to release the strikers.[13]

As the freed miners were making their way home, four of their colleagues were facing similar charges before Captain Spalding in the Nanaimo court and were achieving much less success. After what was becoming a standard objection to the magistrate, tarred with the brush of possible association with the mining company, evidence was taken and the men were committed for trial before a superior court on the charge of unlawful intimidation to which had been added the charge of conspiracy.[14] The men suffered in suspense until the end of May, when, after a trial, confessed by the judge to have contained more contradictory evidence than he had ever heard, they were cleared, in a half-hearted way, by a jury which returned a verdict of "not guilty" with the added comment that they had given the accused the benefit of the doubt in the face of conflicting evidence.[15]

It appeared that Robert Dunsmuir had discovered the "power of the press" and, for one so sure of the rightness of his own position, used it widely to get across his version of the facts to the public. On April 8 the *Colonist* printed another "Card" bearing his signature listing the number of days worked by half his mining force and by each man's name appeared the wages he had received after charges had been deducted for oil and blasting powder. Many of the miners, he claimed, boasted that they only need work six hours daily to make their wages and in addition each man received a free ton of coal monthly. The intention of the card was to show that the miners were getting good wages without the pay increase they were demanding and could do even better if they were prepared to work a full day. The men were carefully referred to as "the miners lately employed at Wellington Colliery."[16]

By coincidence — or perhaps by intention — directly under Dunsmuir's card was a letter signed "A Working Man."[17] He claimed that the strike was called at the instigation of a handful of men and did not have general approval. The men were frightened for "if only one coal miner makes a proposition to stand out for higher wages, whether right or wrong, the whole crowd are bound to follow — each man being afraid of expressing a contrary opinion lest he should be called a 'blackleg'." The leaders of the strike were now "putting the masters to a little trouble and expense'' out of spite, for they know they will not be employed again whatever happens. His advice to the miners was to send back the malcontents to the goldmines from where they came and get back to work.

Another ''Card from Mr. Dunsmuir'' brought to an end his publicity efforts.[18] His first card listing miners' wages was disbelieved, here then was another list on which many of the wages paid were even higher. The average daily wage shown was $3.70. He would have published the wages earlier ''in this strike,'' but feared causing trouble at other mines where he knew the wages were lower. He was only publishing now because ''begging committees'' were going round collecting money from people who did not earn as much as the miners when they were willing to work. The intention was clear: any avenue of help for the miners would be blocked if at all possible. Dunsmuir had made a slight slip by referring to the work stoppage as a strike.

Robert Dunsmuir had also become adept at using the law as a weapon. He was not winning cases, but was demonstrating that anyone foolish enough to cross him could expect trouble. Guilty or not, they would be put through the humiliation, anxiety and expense of a court appearance. He did not hesitate to publish materials bearing on cases before the courts and, strangely, no one seemed to protest — not even when on one occasion he cross-examined a defence witness in court on a charge he had laid. And all the time he continued to sit on the bench as a Justice of the Peace ruling on local misdemeanours.

Although one attempt at bringing in strike-breakers had failed, the company was ready to try again. Thirty-eight miners, recruited in San Francisco, arrived on the *Maude* to begin work in the Wellington Mine. On trying to get to work the next morning, they were ambushed and put to flight by what a newspaper correspondent called ''a numerous body of 'infantry in arms'.''[19] This infantry was made up of miners wives, each carrying a child in her arms and asking if the strike-breakers were prepared to take the bread out of the mouths of women and children? In the words of the eye-witness, ''Thus confronted the new men hesitated to fulfill their engagements, and at length yielded to the persuasive argument of the fair sex and the cries of the children and declined to go to work.''[20] Soon the miners and the supposed strike-breakers were mingling as though they had been friends for years and as on the previous occasion fares back to San Francisco were provided.[21]

Dunsmuir had failed to re-open the mine again, and it seemed that defeat was recognized for no more attempts were made to bring in outsiders. Which is not to say that Dunsmuir, Diggle and Co. were about to give up, or that they had run out of ideas.

One of the great weaknesses of the miners' position was that they occupied houses rented from Dunsmuir and and this provided him with another lever to pull. At the threat of strike the men had been discharged, as far as he was concerned, and so had no right to occupy his houses. Getting an eviction order was no great problem. The *Daily Colonist* reported in mid-March that the magistrate had given the miners until the 28th day of the month to vacate the houses of the Wellington Colliery Company.[22] The miners had nowhere to go. Jobs were scarce and so were houses. They could not go and they should not stay. How could they walk out of their homes carrying whatever belongings they could manage? Where would they put their wives and children? They could have crawled back to Dunsmuir and begged for their old jobs back at any wage rate it pleased him to pay, but they were no less determined than he that they would be on the winning side. Furthermore, there was no certainty that he would take them back. They must stay and fight the evictions. But the law must be obeyed and if legal documents could not move them, then force must. Sheriff Harris and his men arrived to carry out the evictions. Having first tried and failed to get the miners to move out peacefully, Harris ordered his men to evict the tenants forcibly. The men flatly refused to go. The red-faced sheriff then got twelve volunteer bailiffs from the crew of the *Rocket*, a naval vessel in harbour at Dunsmuir's request, and they managed to eject two families before the end of the day. In the evening an impromptu "orchestra" of oilcan beating miners called upon Sheriff Harris and Robert Dunsmuir, and, after serenading them, ended the performance with "three cheers and three groans for the brave men who had turned the women and cripples out of the house."[23] The next day Harris ordered more people and again they refused to go. At this point he was forced to retreat to Victoria to request a stronger force of men to help him carry out his orders.

By winning the battle the miners had lost the war. Dunsmuir had prepared for them a classic trap, and they had fallen into it. Whether or not he regained the houses, for which he had no immediate use, was irrelevant: the evictions might at best force the strikers to give up their resistance and if not then their defiance would set them at odds with the law. He could not lose either way. To the general public the strikers' resistance to the legal authorities raised doubts; the public had been sympathetic, but now questioned whether any group of citizens had the right to defy the laws of the land for any reason. The editor of the Colonist immediately weighed in with his opinion which was that they should "not continue to set the constituted authorities at defiance"[24] for if they did it would be at their own peril. He correctly stated that by their own actions the strikers had raised up against themselves a "new and more powerful adversary than Dunsmuir, Diggle and Company."[25] Dunsmuir no longer needed to fight; others would do it for him and the outcome was certain.

Besides bringing in the force of the law against the strikers, there had been an almost casual use of military personnel; in fact, an illegal use in that no request for assistance had been made by any of the civil authorities empowered to do so. Sheriff Harris on his own initiative had asked for volunteer bailiffs from the naval ship, *Rocket*, and none of the officers of the ship seem to have taken any action to prevent the men under their command from participating in an action aimed at civilians. And why was the same naval vessel used to transport the sheriff and his men to Nanaimo to carry out evictions when at the time of its leaving Victoria there had been no violence as a consequence of the impending evictions? The finger must point to Robert Dunsmuir and his British naval officer partners in the coal trade.

C.F.Houghton favoured simple, straight-forward action and that was why he was in court on an assault charge. Not liking what the editor of a local newspaper had written of him on meeting the miscreant outside the post office in Victoria he had beaten him about the head and face with his stick. Justice Courtney on fining Houghton $50 remarked that it was not at all manly to attack a man unawares: cowardly, in fact.[26] The public scolding would rankle long after the conviction had been quashed by the Supreme Court a month later.[27] It was not amusing for a Lieutenant Colonel in charge of the Militia to be called cowardly in public and Houghton was well aware that many Victorians were sniggering behind his back and enjoying his discomfiture. Action was needed to restore him to his previous position of importance and dignity.

Lt. Col. C.F. Houghton, victor of Wellington, in full-dress uniform.

Action was near. Lt. Colonel Houghton — future husband of Robert Dunsmuir's daughter Marion and commander of the British Columbia Militia — was about to justify his military existence. He had just signed a Militia notice ordering his troops to "proceed to Nanaimo in aid of the civil power."[28] The miners at Wellington were defying the law and what the civil authorities could not make them do the military must. Copies of the order were sent to Companies 1 and 2 of the Corps of Victoria Rifles and to the New Westminster Rifles and the Seymour Artillery on the mainland. The Nanaimo Company of one officer and twelve men was not to be called out as "None of these men were properly trained."[29] The militia notice calling out the troops was dated April 26, 1877 but it was not until two days later that Houghton received a Military Requisition requesting his aid. The requisition, signed by Stipendiary Magistrate Spalding and three Justices of the Peace read:

> Whereas it has been brought to our notice that the miners on strike at the Wellington Colliery, Departure Bay have

> resisted the Sheriff and his officers in the execution of their duties and in our opinion a riot or disturbance will be beyond the powers of the Civil Authorities to suppress or prevent and deal with, We therefore request you to call out such portion of the active militia as you consider necessary for the purpose of aiding the Civil Power and of preventing or suppressing any such anticipated riot or disturbance of the peace.[30]

Lt. Colonel Houghton, it seems, was given to caution, and if there was to be a battle, he intended to be on the winning side. The force he considered necessary was, with the exception of the Nanaimo Company, his entire command. His letter reporting to his superior officer in Ottawa the arrival of the Military Requisition explained that, as there were one hundred men on strike and others sympathetic, he was obliged to take as large a force as possible. He added the further information that the Attorney-General of British Columbia had agreed to pay the cost of the expedition.[31] With the Victoria Rifles onboard, the steamer *Maude* left Victoria for New Westminster to pick up the rest of the troops before leaving for action in Nanaimo. The voyage was uneventful except that off Ladner a house was seen to be burning on the shore and the troops (dashing off to war) stopped in to see if they could be of assistance. It was too late; the house was beyond hope of saving.

On the morning the Victoria Militia arrived in New Westminster, the day's edition of the *Mainland Guardian* reported that on the past Tuesday the Westminster Rifles had elected new officers. All members were advised that another meeting would be held on the following Monday and they should attend.[32] They would be in attendance without a doubt, but not in the way they expected, for with the arrival of Lt. Colonel Houghton on the *Maude* they would be called to the Drill Hall to hear how they should behave on active service. On a Sunday evening, the 51 officers and men of the combined strength of the New Westminster Rifles and the Seymour Artillery embarked and prepared to leave with the dawning of the next day. The New Westminster troops were allocated a room "about fourteen feet square with a damp floor underneath"[33] and there, as night fell, they made their simple preparations for sleep: mattresses down, blankets spread and boots off. It was not easy to sleep. There were those with a

Militia-carrier Maude, *moored below Nanaimo's Bastion.*

sense of the dramatic, rather overblown for the circumstances, who were determined that, if life was going to be short, it should also be merry and so joked and sang to the annoyance of the more staid. As time passed even the most boisterous calmed down and most of the men slept, disturbed only by the sound of the sentry's boots thudding on the deck overhead. The New Westminster troops were not alone in their discomfort for when at the first light on Monday morning the *Maude* headed down the Fraser River enroute for Nanaimo, Houghton had onboard 101 men and 8 officers stowed away in every nook and cranny, "even to the tables in the saloon which were densely packed with men both on top and underneath."[34] There was one armed man for each striking miner and a few left over.

The temporary soldiers were soon reminded that "war is hell" for with the serving of breakfast a breeze sprang up and the *Maude* began to dance on the waves. The reporter for the *Mainland Guardian*, who seems to have been a militiaman with a strong sense of humour as well as a strong stomach, gloated that "it was gratifying to see the liberal manner in which our new fledged superiors shared their last mouthful with Neptune, God of the Sea."[35] Seasick or not, when the ship reached Departure Bay they would have to be ready for action.

At eleven o'clock when the militia stepped ashore each man carried twenty rounds of ammunition for use in the coming encounter. Robert Dunsmuir and Magistrate Spalding met the troops and led them to a train of empty coal trucks in which they travelled over the company's own lines to Wellington. Captain Harris, the sheriff, making yet another attempt at evicting the striking miners, travelled with them. The unlawful possession of the Dunsmuir houses had gone on long enough. This time the force of the law and the armed strength of the militia would ensure success.

The strikers stood in a sullen group as the train arrived in Wellington. Only an hour before they had attended the funeral of a small boy killed in an accident. The whole village was in mourning. And now the militia was here. They knew that the militia had been called out but never expected to see troops actually arrive. It was too much to grasp — armed soldiers in Wellington to be used against civilians: to be used against them! All the strikers could do was to manage a few cat-calls and jeers. "My G-d what's them?" ". . . ain't there a lot of 'em too!" ". . . a dose or two of buckshot would clear 'em all out,"[36] and so on. The only response from the militia was good-natured laughter and, tired of wasting their efforts, the miners wandered off to see what the bailiffs were up to.

On the first afternoon of bailiff action several families were evicted. However, the *British Colonist* reporter, "In the Tented Field"[37] as his headline described it, claimed "there were very few pathetic scenes." The weeping of the women, as he saw it, came not so much from the misery of being ejected as from rage and vexation, while the men acted as though they were simply obeying the instructions of their legal advisor rather than behaving as he would expect of men surrendering to ejection by overwhelming forces. Some were resisting "to get square with Dunsmuir for the money they claimed to have lost by being given short weight on coal they had mined for him." Whether or not the correspondent's perceptions were accurate is open to conjecture, but there was, in fact, very little violence. A few houses had been broken into without opposition, except that the old women heaped abuse on the Dunsmuirs. An old lady had been allowed to stay in her house on the promise that she would leave peacefully when her son returned to help her to move. A miner named Baker used threatening language to the sheriff when his house was broken into. He was removed by his friends, only to be found later, back in the house, struggling with a bailiff for possession of an axe. Baker was arrested and struggled violently until four constables managed to subdue him enough to get handcuffs on his wrists, and even then he asked several times for the handcuffs to be removed so that he "might get satisfaction from the bailiff" who, he claimed, had struck first. Only when told that he would be released later in the day did he gradually calm down. Three other miners were arrested that day and, curiously, that was for resistance to the sheriff two weeks before. It seemed that the sheriff had been too prudent to go in and get them on the day of their resistance and, anticipating returning later with a larger force, had left the arrest to a safer time.

The work of eviction was slow, for not only had the belongings of those ejected to be carried from the houses, but they also had to be carried down to the railway line and loaded onto the coal trucks provided by Dunsmuir for removal from the company property. There was insufficient time to complete the evictions on the first day and the bailiffs and the militia were withdrawn, with the exception of a handful of men left to guard the mine premises — premises which had been perfectly safe and undamaged throughout the time the strike had already been underway.

On the second and final day of the evictions, Lt. Colonel Houghton did not consider it necessary to have "more than one company on the ground at the same time" and only once was assistance requested by the bailiffs, at which time a single section of a single company advanced to "within some twenty-five to thirty yards of the operation."[38] The soldiers, being only part-timers and having home responsibilities, were anxious to be off and by late in the afternoon were back onboard the *Maude* for the return journey. Early next morning the ship sailed for New Westminster and after delivering the mainland contingent headed for Victoria. At the mouth of the Fraser River headwinds and a strong current caused a delay at the sandbars and then within fifteen miles of Victoria the ship ran onto rocks some distance from the shore. In the minor panic ensuing, boats were launched and headed for the shore but most of the men stayed where they were and were picked up at first light and taken on to their destination.

The military operation was at an end, the militiamen, for the most part, having seen their first and last action — if action it was. Was it a serious incident or just a brief and unexpected holiday lark? The actions of Private Perce left little doubt. His active military experience had been spent "fraternising with the enemy" in that "when on duty under arms, left his Coy. contrary to orders, in the company of a miner . . . returned later drunk and incapable. Reprimanded." Perce was not very impressed by the reprimand for he then managed to get drunk again on the voyage back to Victoria, and as a result, before his comrades in arms, was "ignominiously dismissed from the service."[39]

The editor of the *Daily British Colonist* was sorry to see women and children turned out of their homes and regretted that such an extreme step as the use of the militia had been necessary, but otherwise he considered the shock to the law of continued defiance would have been fatal to law and order. In sending in soldiers the government had acted in a manner "worthy of a vigorous administration."[40] In his opinion, the ringleaders who were in custody were dependent for the severity of their punishment on the future actions of their fellow strikers — a curious view of the law when punishment depends not on what you have done but rather on what someone else might do in the future. His advice to the strikers was to get back to work at the pay rate offered or make way for others. Outside Vancouver Island the incident made scarcely a ripple. The *Mainland Guardian* in its main editorial on the day the militia returned from Wellington dealt with the prices farmers were being paid for grain.[41] A week later the correspondent of the same newspaper, writing from Victoria, was lamenting that "Were it not for the war news from the East the local papers here would have been most uninteresting."[42] He did mention in passing that the Wellington miners arrested for intimidating the sheriff were to be tried.

As a military affair the operation at Wellington bears the stamp of a comic opera, but there was another side. The threat of the military had been real, as real as the bullets issued to the troops before landing. Under the guise of maintaining law and order, Robert Dunsmuir had manipulated the situation to cause the civil authorities to fight what was in fact his private battle with the strikers. He had made very clear his point that he could not and would not be intimidated by the actions of his employees,. and had shown not the slightest compunction in urging military action against his workers, some of whom were one time colleagues from the days when his situation had been less elevated. The main importance of the incident lies not in the degree of force used but in its symbolism; in the resentment it bred in the miners, a continued resentment that would project into the future with greater effect as it became swollen with myth. It was, and would continue to be for the miners, a perfect illustration of the power and ruthlessness of employers when in any way thwarted by the workers attempts to get what they considered to be simple justice and a living wage. Further, in that the government of the province had paid the expenses of the military action, the miners now believed they had proof that politicians could always be expected to ally themselves with the rich at the expense of the poor and weak.

The rumblings and repercussions of the evictions and of the militia's part in them echoed around Nanaimo for a long time. The Grand Jury of the town regretted that a "serious stigma" had been cast upon the district by military intervention and stated that the cause had been the incompetence of the sheriff in the execution of his duty.[43] Sir Matthew Begbie, to whom the complaint was addressed, wasted no words in assuring the Grand Jury of Nanaimo that, in his opinion, the cause was the unlawful act of the miners in retaining possession of the Dunsmuir houses in defiance of the court order, and that occurred before the sheriff had even appeared on the scene — "no want of judgement on his part could make their wrong right." [44] In the local communities petitions were circulated for the removal from office of both Captain Spalding, the magistrate, and Captain Harris, the sheriff, but to no avail. The three strikers, Hoggan, Thompson and Phillips who were arrested on the first day of the military action for offences committed previously, were given a hearing and sent to trial by the Supreme Court. To the expected charge of obstructing the execution of a court ordered eviction was added a charge of conspiring to raise wages — a charge that surely could have been brought against each one of the strikers. Eventually Alex Hoggan served a two month jail sentence for his resistance to the sheriff and seems to have been the only participant in the various strike-related cases to have attained such a distinction.

After all this the strike was still not over, but the end was near. By mid-May Dunsmuir was posting notices asking those of his former employees who were willing to return to work at the rate of $1 per ton of coal raised to send their names to him. A slight crack in the facade of the strikers solidarity appeared when a dozen men did so and were re-engaged, only to be prevented from working by the rest of the strikers who were still determined that the mine should remain closed. The effect of this action was to lose the miners whatever shred of public support they had remaining to them. In an editorial, which no doubt fairly accurately reflected the feelings of many people, the editor of the *Colonist* expressed his revulsion at the miners preventing others from working and urged strong action. He pointed out that the strikers had already had the military used against them and some of them had been turned out of their homes and now, when given a chance to go back to work they were still refusing. After all that had happened, they were still breaking the law and at a time when they should be "moderate in their demands and conciliatory in conduct."[45] He went on to state that, if the managers decided to bring in outsiders to open the mine, they would have every right to do so, and, if protection was needed, then a well organized body of police should be sent to Wellington to give whatever protection might be called for.

Dunsmuir had every intention of opening the mine again now that public opinion was behind him. Forty men from San Francisco were landed towards the end of May and accommodated in the empty cottages from which the miners and their families had earlier been evicted.[46] In spite of the efforts of the strikers, the new men went to work. Sensing that the end was near, so did thirty of the strikers.[47] The will of the strikers was broken and they all applied for work, stating that they "had been badly advised and (were) deeply regretting the course they had been led to pursue."[48] Not all of them were taken back, however. While Alexander Dunsmuir freely gave work to strangers, he "selected 25 of the least obnoxious"[49] of the previous employees, a phrase implying that all were to some extent obnoxious. With the arrival of thirty-one Italians from outside the area, the personnel needs of the mine were met both by day and by night and very shortly three shifts were at work building up a stockpile of coal for export.[50]

The four-month strike was at an end. The strikers had lost wages for all that time; some had lost their homes and their employment. Those lucky enough to be re-engaged received the same pay rate they had used as a justification for striking. After enduring great hardship they had been thoroughly and decisively beaten. But that was not the end to it. The men had organized a union, they had been given a deep sense of grievance, and they had come to realize that the gulf between capital and labour was so deep as to be beyond bridging by discussion and persuasion. If they wanted higher wages and better working conditions it seemed that they would only come through applying even greater pressure. In Robert Dunsmuir and his sons they had found a focus for their enmity.

Throughout the strike Dunsmuir had not wavered from his original position by one iota, and he had not hesitated to use every available means to get his own way. His armoury of weapons was wide and was effectively used against selected individuals rather than against the mass of strikers. Anyone standing out from the crowd was immediately hit, either by newspaper ridicule, court action or eviction, and certainly by having all hope of ever being employed again in Dunsmuir mines completely removed. The message was simple and clear — cross me and you are in trouble. The trouble was sufficient to cause all but the bravest, or the most foolhardy, to pause for reflection before taking any action against him or even publicly proposing action. Robert Dunsmuir's use of newspapers and the law was effective in that he was able to manipulate events so that the strikers always appeared in the worst possible light and so were bound to lose public sympathy. To the end of his life he would refuse to recognize unions, or any other permanent organization of miners. Shortly before his death he said that, while he was always ready to receive a communication from his men delivered by a committee, he "certainly could never recognize a standing grievance committee upon any consideration."[51] Whatever actions Dunsmuir took during the strike, he took, in his view, in the legitimate defence of his rights. He believed, and many agreed, that he was in a position of exact equality with his men. They lived under the same rules. He was no more forced to employ a man at the wages the man demanded than the man was forced to work for him at the wages he was prepared to offer. It was a miner's right to ask for higher wages and his right to refuse to pay them. If he was not satisfied with a man's work, he would dismiss him in much the same way that a man dissatisfied with him as an employer could choose to leave his employ. What could be fairer?

Fair, indeed! It was a surface equality at best hiding an ocean of inequality. A man dismissed immediately lost his ability to maintain his family and at the same time would be obliged to move out of his company-owned house. Could a man in such a position have any real possibility of leaving an employer because he was unhappy with the employer? Hardly. The most damage the employer could ever suffer was a temporary inconvenience until the man was replaced. The claim of equality was repeated in letters to the editor and even by the Chief Justice of British Columbia in remarks addressed to the Grand Jury of Nanaimo. Of all the things the striking miners had to suffer, this pious myth of equality must have been the hardest to bear. There is little wonder cynics have remarked that some people are more equal than others.

6. Disaster

Mining can never be safe. When man burrows into the earth, removing as he goes the natural support of the overlying rocks, there must surely be roof falls. When he blasts the coal to make its removal easier, there must be fine coal dust to coat not only the body but also the interior of the lungs. When equipment is used in confined and ill-lit spaces, there will be danger. Should gas seep from the rock and gather in invisible pools, sparks and open flames will cause explosions, or perhaps the gas will insidiously and silently cut off the life-giving air, and consciousness and life will gently slip away. When the different languages spoken by those labouring underground cause confusion, fatal errors become not only possible but probable.

There was no doubt that the mines of Vancouver Island were among the most dangerous in the world. They were wet and gassy, the rock strata broken and uncertain, and in places the seams reached far out under the sea. A Nanaimo miner had a three times greater chance of dying in the mine than had a miner anywhere else in the British Empire. And the figures for those dying did not measure the pain and suffering of men who were incapacitated by accidents which occurred in a never ending stream. Men lingered and died from their injuries without their deaths ever being recorded as the direct result of mining accidents. Without the sudden catastrophe of an accident, men also sickened and died from breathing coal dust for years on end. Often the owners were resistant to safety regulations, seeing any mine legislation as an unjustified intrusion into their business affairs and, as most safety measures had a cost, a limitation on their profits. In British

Columbia, then on the distant fringe of the settled world, mining methods lagged far behind those of other countries. To give only one example, the miners' safety lamp, which was invented in 1815 and spread rapidly through the mines of the British Isles, was not in use in the mines of Nanaimo forty years later where the open flames of fish-oil lamps were still inviting disaster. The miners themselves were not free of fault in their own attitudes to safety and some were inclined to take short cuts and create dangers which were avoidable. As they often worked on piecework and had to hew coal rapidly to make a decent living it could be argued that the system rather than they was at fault; the effect was the same. The problem was far from simple: the geological and chemical nature of the coalfield could not be changed, the attitudes of the owners and miners possibly could.

Even in the wide open days of laissez-faire economics it was not possible for governments to entirely escape responsibility for working conditions. As the number of miners grew and the importance of their votes thus became greater, the politicians were made aware of mine safety as a concern of their constituents. The mine owners might not like regulation, but governments work through regulation. The miners pressed for a version of the Coal Mine Regulation Act passed by the British Parliament. They were heard, but it was not until 1877 that British Columbia passed from having no safety regulations other than what individual mine owners saw fit to permit to having an act said in a debate in the Provincial Legislature to be the "English Act adapted to this locality." Having been worked on by a delegation of miners for "nearly a whole day," the Act was considered by a mining expert to be "almost perfect."[1]

It was at the opening of the Second Provincial Legislature that the Lieutenant Governor of the Province gave notice in the Speech from the Throne that measures would be proposed which when applied in the coal mines would aid in "decreasing the danger of calamitous accidents."[2] The issue was considered "non-political" and, when the Hon. Mr. Elliot, Premier of the Province, proposed the adoption of the bill, Mr. Walkem, Leader of the Opposition, immediately rose to second the motion. It could hardly have been otherwise, for the matter had been first raised in the 1875 election campaign, which had been won by Walkem, and he, as Premier, had sponsored a similar bill which had been shelved when his government fell. Apart from its importance to mine safety, the bill specified who could work in the mines and for how long. Boys of thirteen years of age, but not yet fourteen, were permitted to work on only five days of the week and for a maximum of six hours daily; presumably on becoming fourteen they became men as far as working hours went. Women were not permitted to work underground but could be employed at the surface, a topic which caused the Premier to express concern that women's rights were being interfered with and caused another member of the house, Mr. Smith, to observe that:

> The moment you tell a women she mustn't do a thing she wants to do it. (Laughter) If you say women shan't work in the mines they will drop their pianos and crochet-work and crowd into the mines three deep looking for work. (Great laughter) Don't tell a woman she shan't do anything if you want her to do it. Tell her she may and then may be she won't. (Roars of laughter).[3]

The immediate reaction of the mine owners and managers on hearing that a Coal Mines Regulations Act was contemplated was to send a petition asking the Government not to legislate with undue haste, as such an important question was worthy of study by a select committee or a commission which would first enquire into the necessity for such a measure. It seemed to them that the proposed act would work oppressively and unjustly upon colliery owners, and cause them vexation and loss without benefit to anyone.[4] The bill passed.

Having a Mines Safety Act is not the same as having those concerned comply with its terms, as Edward Lawlor Prior, the first Inspector of Mines, was soon to find. His immediate problem on appointment was in getting the Act, and himself, taken seriously. The act was passed in mid-August. By the first day of November Prior had already inspected several mines, and, by means of letters, was beginning his uphill battle with the mine owners and managers. To Manager Bryden of the Vancouver Coal Company he complained of poor ventilation and pointed out that "no Chinaman is allowed to be in a responsible position such as attending a ventilating furnace etc. whether he can speak English or not" — the bill had built-in racial discrimination. He condemned the common use of iron tampers and needles, and recommended that they be replaced with similar tools made of copper which would prevent sparks and cut down the explosion hazard. In April of the next year he was still complaining that the substitution had not been made at the mines of Bryden and Dunsmuir. They were given until the twentieth of the month to comply. Written accident reports as required under the act were largely ignored. Prior got most of his information from the newspapers and was then obliged to follow up with the recalcitrant owners who, as often as not, had to be sent numerous reminders before the report was finally sent to him. "Special Rules" for the operation of particular mines were difficult to get from the owners and, when they came, he had to ask the managers to assure him that they had been posted where the men could see them. Even so he remained unsure that his request had been carried out and sent two hundred copies of the rules to Bryden and Dunsmuir with the request that they be handed out to miners asking for a copy. The owners were not open and blatant in their opposition to the Act, but they certainly did not go out of their way to comply with its terms. Prior was adamant that the safety rules of the act should be strictly applied, but, knowing there was a cost to the owners, in a fair and even manner. To the Minister of Mines he wrote:

> There can be no doubt that if the act is to do any good it must be strictly enforced, for if not it causes injustice to those owners who do their utmost to comply with the act, and thereby incur a certain amount of expense, which is avoided by those who wilfully ignore the same.

But it was not only the managers and owners that caused the Inspector concern, the miners, who had demanded safety regulations in the first place, he found to be "constantly breaking the rules and thereby causing an incessant source of danger."[5]

Regulations are a guide, not a guarantee, and some people do not wish to be guided. Of all the men down a mine it takes only one to create a disaster; for all the supervision, one piece of 'blindness', accidental or deliberate, can blow up a mine; for all the safety precautions that are in force, one omitted, for the sake of cost or convenience, can lead to a sudden and frightful death. Coal mines are not forgiving.

Tuesday, May 3, 1887 seemed no different to any other day as underground workers straggled off in loose groups to begin their afternoon shift in the Vancouver Coal Company's Number 1 mine on the Nanaimo waterfront. They were a mixed lot: old George Bowden, who had come from the rocky coves of Cornwall, was seventy-one years of age and still putting in a shift underground; somewhere in the throng walked Andrew Hunter, mule driver, and Michael Lyons, both aged fifteen. James Lyons, father of Michael, was there also. Not only was there a mixture of ages but also a wide range of national origins. Nanaimo had obviously changed; those of British descent were still in the majority but now there were also Swedes, Americans, Italians, and Germans, and to one side, distinct and separate, walked more than fifty Chinamen. With a last look at the sun and the open sky, they entered the cage a group at a time and dropped rapidly into the black depths of the earth.

Later, almost halfway through the shift, the mine "fired." Miller, the engineer in charge of the fan engine above the ventilating air shaft on the surface, was suddenly blown through the door of the fan house by the force of an underground explosion venting through the shaft. As he dazedly rose to his feet, there was a second and larger explosion which shot debris from the mine below hundreds of feet into the air. Then flames roared out of the shaft, soon reducing the fan house to a charred wreck. In Nanaimo, poised always on the brink of disaster and where any unusual noise caused people to look at each other with eyes filled with dread, there was no doubt as to what the rumbling and vibrating of the earth meant. The mine whistle began to wail its eerie, continuous note, warning of things to come and appealing inarticulately for help. Even as it started to wail men were already running towards the pithead to join the rescue parties. The simple and undramatic tradition of the miners was being obeyed: when men are in trouble down the mine they must be got out. Some of the men had sons on the afternoon shift, some brothers, some fathers. No one held back — least of all the managers and oversmen — the miners might kick against "boss's men" in the normal course of events, but they well knew that they would be the first to put their own lives in jeopardy in a mine rescue. No distinctions would be made; managers and miners from all the collieries of the area would be there to do whatever they could.

The first party to go down the mine was led by William McGregor, the mine manager. Soon the wheels of the headframe began to spin as the cage was sent back to the surface. The crowd surged forward as seven covered and still shapes were unloaded; four were Chinese killed outright by the force of the explosion and the other three, killed by afterdamp, were white miners. Again a cage reached the surface. This time, although they were burned and had broken limbs, the six passengers were alive. Later in the evening Richard Gibson, oversman of the mine and the Mayor of Nanaimo, was found by a rescue party dazed and alone, marked only by a few bruises. He was unable to say what had happened or how he had come to be by the underground stables where he was found. All the men of the afternoon shift that would survive were now out of the mine but the rescuers were not to know that and the search would continue until every man had been accounted for. To the names of the dead would be added that of Samuel Hudson of Wellington, who died of the effects of afterdamp during the rescue attempts of the first evening. He had upheld the tradition of the mines at the cost of his own life.

There were 148 men still in the mine. At the foot of the air shaft the fire still raged, making it impossible for the rescuers to penetrate into the deeper and more distant workings. The fan which delivered fresh air to the underground workings was a mass of wreckage. For those still alive, the greatest danger was the spread of the gas, afterdamp, which had already claimed the life of Hudson. But until the fire was controlled most of the trapped men were beyond reach. Throughout the first night volunteers from ships in the harbour poured water on the fire in an attempt to douse the flames. Miners from all the collieries in the area, led by Bryden, Little and Chandler of the Dunsmuir mines and by William McGregor of the stricken mine, risked their lives in the hot and gassy underground passages. They were accompanied by Archibald Dick, the provincial inspector of mines, and by the former and first inspector, Edward Prior, who as a Member of Parliament had by then no direct link with the mines, but believed that, as he had spent years in the pits with many of the missing men, it was natural that he would give whatever help he could. Progress was slow and with each passing hour the chance of finding survivors lessened.[6] At the pit head crowds surged forward each time the cage came to the surface hoping to see rescued men; women with children clinging to them searched the faces of those leaving the cage with a desperation born of fear. At first they saw only the weary, grimed faces of frustrated rescuers. Then the still, covered forms of the dead began to arrive at the surface to be identified and taken to the Old School building where they would be claimed by relatives. The misery of the scene was made worse by the streaming rain; even the weather was against them it seemed. The Merryweather steam fire engine pumping water to the mine from the beach blew a valve due to being frequently clogged by seaweed stirred up in the bay by the gale force wind. This left only a hand-operated pump to deal with what remained of the underground fire.

The explosion happened at Number 1 mine on Tuesday evening. It was not until Sunday that the fire was put out and bodies began to be recovered from the distant parts of the mine. Some men had obviously died immediately from the force of the explosion, but others had lingered until claimed by the silent advance of afterdamp. Three quarters of a mile in along the No.1 level, were found the bodies of twenty-five white miners and ten Chinese assistants. Among the dead was David Ellis; his luck had run out after surviving three previous underground explosions by building air-tight barricades against the approaching afterdamp. The barricade was in position but this time it failed Ellis. It also failed his twenty-year old son who was found with him; the insidious gas had worked its way to them from the rear. The bodies appeared to be in a natural sleep and seemed so unaffected by the days they had been entombed that a rumour grew that the men had lived until shortly before the rescue party had broken through to them. The evidence found by the rescuers suggested otherwise. Scratched in the whitish mould growing on mine timbers were the last messages of two miners:[7]

> 13 hours in misery — JOHN STEVENS
> 1, 2, 3, 5 O'CLOCK — W.BONE

If the messages had been prompted by a feeling of impending death it seems likely that, as the explosion took place at six in the evening, the times scratched on the mine timber by William Bone referred to the morning of the next day. Stevens' last message was probably written two hours later at about seven in the morning. As the bodies showed no sign of fasting and untouched food was found in a dinner bucket, it seems that the end was not much later than the messages suggest. On the mould of uprights near the dead Chinese were scratched Chinese characters which the rescuers were unable to decipher.

When at last it was all over there were 148 bodies. There was old George Bowden, aged 71; the two fifteen-year-olds, Andrew Hunter and Michael Lyons; there was James Lyons, the father of Michael. There were men who had been the coal-mining pioneers, Archibald Muir of the original Fort Rupert party; the Brierley Hill pioneer John Meakin and, of the Nanaimo-raised generation, his son John and a son each of the Bevilockway and Biggs families.[8] There were men with the familiar names of the British Isles and men with the more recent and less familiar names of Germany and Sweden and Italy. There were forty-six new widows in Nanaimo; 126 children of various ages had been deprived of a father. But that was only a part of the story: no one knew, or possibly gave a thought to, how many widows and fatherless children had been created in far off China. All fifty two of the Chinese workers had died. No Chinese names appeared in the newspaper lists reporting casualties and of the Chinese bodies recovered it was simply said that they were "brought to the surface and claimed by relatives." When, nearing the end of the rescue operation, a *Nanaimo Free Press* report was headed "Six Bodies remaining in stalls off the Airway," only at the very end of the report was there any intimation that the bodies of seven Chinese also remained to be found in that location.[9] The Chinese, apart in life, were no less apart in death.

What a *Colonist* reporter oddly called "a sad carnival of death" reached beyond the direct victims. Mary Culligan, a friend of young Michael Lyons and engaged to be married to a young man killed in the explosion, "incautiously visited the school house where so many bodies were laying. The shock brought on convulsions, and although all the medical men were called in she died at 1 o'clock on Monday morning." A few months earlier her father had been seriously injured in the same mine. An extreme, but apt, example of how coal and the hidden workings of the mine reached into the homes of the miners' families. The mind-numbing grief of the period would not quickly be forgotten but life went on. Death and and triviality existed cheek by jowl. Side by side with the lists of the dead, newspapers carried advertisements of drug stores and notices of coming auctions. The Nanaimo School Trustees, a week after the explosion that had created a swarm of orphans in its schools, discussed as an item of important business a resolution condemning a rule allowing teachers three days "holiday" to visit other schools.

Life went on; schools re-opened. A foundation was prepared for a new fan house so that No.1 shaft could be put back into use; and the remaining miners went back down the mine. With what trepidation they first entered the cage to go below ground where so many of their friends and relatives had died is not recorded. How easily they were startled by underground sounds they would once have considered normal and what the feelings of their wives and families were as they set out for the mine each day we cannot tell. Life had to go on and the mine was life; however they might talk about leaving the mine, they would stay — they had little alternative.

A coroner's inquest was called to inquire into the explosion. Financial matters were dealt with first. The foreman of the jury wished to know if the jury would be paid for their expenses and lost time. The Minister of Mines, with a politician's flair for leaving loopholes, suggested that if the matter extended over a "considerable period" of time the government would "favourably consider assistance in meeting extra-ordinary expenses." The Corporation of the City of Nanaimo showed no great willingness to pay the costs of the inquest as it was not clear that the deaths had taken place as the result of an accident within the limits of the city. Refusing to concede that the government should meet the costs, the Minister of Mines replied that was a legal point for future decision. The jury found that the explosion was caused by the firing of a badly planted charge which ignited gas. The initial blast was intensified by the addition of coal dust. They were satisfied that the ventilation system supplied enough air, but pointed out that gas was still able to collect in hollows in the roof. Their main criticism of the operation of the mine was that, although coal dust was a known factor in colliery explosions and the mine was known to be dusty, there had been no proper precautions to minimize its effects. They were not, however, prepared to charge criminal negligence.[10] The basic causes were clear: a white miner had improperly placed a charge and its explosive effects had been intensified by conditions the management had failed to see. It was simple enough — but it was not believed.

The miners disbelieved the findings of the jury because they did not fit in with their own fixed opinions. The jury had made no mention of the Chinese underground workers. The white miners were convinced that the disaster had been caused by the Chinese. Even before the last of the bodies had been recovered, placards advising of a miners' meeting appeared. When it took place a committee was set up having as one of its tasks seeing what could be done about barring the Chinese from underground employment.[11] As a result of the miners' activities two petitions were sent to the provincial government requesting that amendments to the Mining act should be made following recommendations made by the miners of the Nanaimo district and that particular attention should be given to the clause relating to the withdrawal of Chinese from underground work. The presence of the Chinese, they were convinced, was a major contribution to the unsafe condition of the mines and their removal was necessary to the personal safety of the miners. Even though there was little direct evidence that this was so, and no evidence at all that the recent disaster was in any way the fault of the Chinese, the first petition was signed by four hundred miners. This was followed by a second petition, almost identical in wording, signed by over three thousand citizens of the province.[12] But the miners no doubt knew better than to expect quick legislative action. The legislature met annually for the two months of February and March with a variable extension into January and April and it was not due to meet again until January of 1888.

In distant London the absentee landlords of the Vancouver Coal Mining and Land Company met to hold their annual general meeting. Mr. John Galsworthy, the chairman of the company, explained why, in spite of a net profit, no dividend would be paid. In his concluding remarks he referred to a telegram giving news of a serious explosion in the No.1 Shaft at the Nanaimo Colliery. The seconder of the motion adopting the report commented on the hopeful outlook of the company, but regretted the explosion.[13]

The shattering accident at the Number 1 Shaft in Nanaimo made the men connected with coal mining extremely sensitive on matters of safety; the Inspector of Mines was no doubt unusually thorough in his inspections of mines in the summer of 1887. Praise for mine safety must surely have been hard to earn and yet praise was what Archibald Dick had for Dunsmuir's No.5 Pit at Wellington. Dick thought this to be the ideal mine and, in spite of what miners often claimed about the parsimony of mine owners, one in which no expense had been spared to make the works safe. The coal was worked by only one shift daily and this gave the mine plenty of time to cool down and time for any slight accumulation of gas to dissipate before the next shift. Even without the resting time it seemed unlikely that gas could gather in sufficient quantitites to be a danger when the ventilation system was so efficient that men had been known to complain of the fan causing such a breeze that it was sometimes difficult to keep a light lit. The inspector found no dust in the mine and concluded that it was safe from anything that could cause an accident.[14] Unfortunately, Inspector Dick was wrong. Six months after making his report, he was explaining to John Robson, Minister of Mines of British Columbia, that there had been an explosion in No.5 Pit and of the one hundred and seventy men men caught below ground sixty-eight had no hope of survival.[15]

A man walking to the Wellington railway station shortly before nine o'clock on Tuesday, January 24, 1888, reported hearing a noise like a cannon and then seeing an immense plume of dust and black smoke towering over the airshaft and fan house of No.5 Pit. After a few minutes the plume subsided and all that could be seen to suggest that anything unusual had happened was that the surface of the newly fallen snow around the fan house was inky with fallen coal dust. As the eye-witness left for Victoria he saw scores of men, women and children streaming towards the pit mouth, those with friends and relatives in the mine crying and lamenting their loss.[16] Below the surface the survivors of a violent explosion, deafened and dazed, groped their way through the pitch black mine towards the main shaft where they expected to find the cage waiting to lift them to safety. It was far from certain that they would ever reach the shaft as the force of the underground explosion venting through the airshaft had stopped the fan and circulation of air in the mine had been halted; the danger now was from the gas which was already reaching for their throats as they struggled on. Many lost consciousness and fell into what was usually the final sleep before death. Then, as if by a miracle, the air started to move again and fallen men stirred, rose and staggered on. But it was not a miracle. On the surface John Bryden, experienced and knowing exactly what the priorities were, had directed feverish efforts to repair the fan and had succeeded in getting the air moving again. In Victoria, Robert Dunsmuir on hearing of the disaster immediately arranged a special train to take him to Wellington to direct the rescue operation. Offers of help flooded in. Samuel Robins of the Vancouver Coal Company offered to close down his Southfield mine and send all the men to Wellington to help in the rescue operation. His offer was declined for, although the miners remaining alive were in a position of extreme discomfort, it seemed that with the air flowing strongly and in the absence of fire they were in no immediate danger.

As the men underground neared the foot of the main shaft they found the going harder for the explosion had shattered the timbering. The main shaft was blocked by debris and the hoisting gear damaged making it impossible for the cage to descend to the bottom of the shaft. Three men, wanting to be out of the mine at any cost, clawed their way upward through thirty feet of broken framing and arrived at the surface exhausted after climbing hand over hand up 150 feet of wire rope. Eventually about ninety men collected at the bottom of the shaft, anxiously awaiting recue as the men above struggled to clear the debris from the shaft and get the cage operating. Not until the next day was it possible to bring up the survivors. But there were many men in the immediate vicinity of the explosion who did not survive. As their bodies were brought to the surface, the skies wept heavy rain on the sad scene.[17]

The newspapers were not slow to supply gory details. It was obvious to the recovery parties that the explosion had been concentrated and violent, but mercifully restricted to a relatively small part of the mine. Anyone in the direct path of the blast had not had a chance of survival, but the rest of the underground shift had been in danger only from afterdamp. In the explosion's path bodies were strewn about, some stripped completely of clothing by the force of the blast, others were gashed about the head and face, some burned around the middle yet with lower limbs free of any damage. Elisha Davis was recognized only by the colour of his shirt for when his mangled trunk was found it was a shapeless mass with most of the head blown off. One of his legs, torn from its socket, was found forty yards away. The father of Elisha Davis was also working in the mine and was one of the first to be rescued. One of the dead had gone down the mine that morning for the first time. Another had been working underground for only two days.

As the bodies of the white miners were recovered, they were taken to the carpenter's shop and stripped of their tattered clothing. Each was washed to remove as much as possible of the effects of the explosion and dressed in a black suit, with a white shirt, collar and tie and placed in a handsome coffin. The coffins were taken to the public hall in Wellington to lie there until Friday morning when the dead would be taken by train to Nanaimo for burial. The dead Chinese, known by a number rather than by name, were washed and dressed by their friends and placed in rough coffins painted black. Each Chinese body was carefully searched and any money or valuables found were parcelled and labelled for future claim by proven relatives. It was reported that some of the Chinese carried as much as $600 on them while two had gold watches and another carried several ounces of gold dust. Living, as they had, in squalid, over-crowded conditions, they probably had no alternative but to carry their valuables with them always. At the Nanaimo disaster of the previous year it was said that some Chinese had claimed bodies, removed valuables from them and then abandoned the bodies for burial by public charity. True or not, no chances were being taken this time. While reporting respectfully on the preparation of the white miners for burial, the *Colonist* reporter wrote of the scene in the blacksmith's shop where the Chinese bodies lay, "While the preparation of the dead was proceeding in every portion of the building, the place inside and out was crowded with celestials and for one unaccustomed to such a scene, the revelry of death was a horrifying spectacle."[18]

Thirty-one white miners had died. The number of Chinese lives lost is difficult to determine: in his original report Mines Inspector Dick suggested thirty-eight, however his final report listed only twenty-nine. Newspaper reports gave the number as thirty-nine and forty-one. No one knew why so many men had died in a mine said to be the safest on the island and one in which men would enthusiastically choose to work if given the choice. But, as always, there were those willing to lay blame for the accident the day it happened.

Robert Beaven, at a political meeting held in Victoria on the evening of the day on which the accident occurred, placed the blame squarely on Robert Dunsmuir leaving the editor of the *Colonist* to ask his "humane hearted" and "right feeling" readers if they could believe that

> advantage was taken by Mr. Beaven last night of the explosion to use it for political purposes . . . a person who writes honorable to his name, publicly disgraced himself by a gross and unwarranted attack upon Mr.Dunsmuir . . . he

> charged that the loss of life both in the Wellington and Nanaimo colliery disasters was largely due to the parsimony of Mr. Dunsmuir: that he had refused his assent to a bill introduced in the House providing for safeguards such as would render calamities of this nature comparatively impossible, and all because it would take a little more money out of his pocket. Robert Dunsmuir, one of the kindest hearted, most liberal, one of the most sensitive natures alive, was painted in the darkest colors as a human monster, who for the sake of a few dollars, did not render sufficient protection to the lives of his employees.[19]

In spite of what Beaven had to say it was the editor's opinion that there was widespread sympathy for Mr. Dunsmuir who at the time of writing was in Wellington to relieve the anguish and soothe the sorrows of the bereaved. Some support for the mine owner came from a source some might have considered unexpected and was contained in a letter printed in the *Colonist* under the heading of "A Miner Speaks." The miner, who professed himself to be disgusted with Beaven's accusation, wrote:

> I have worked for Dunsmuir off and on for seven years, and I know my chums will back me up when I say that there never was a man who tries harder to keep the pits in safe condition than the "old man." Money is never wanting for any such purpose, and he is always ready to listen to anything a man working in the pit may suggest. Is it likely that he would begrudge a few dollars a day if by spending them he could save an explosion. Every explosion costs him from $10,000 to $50,000, and all who know him at all know him to be one of the tenderest-hearted men alive when there is any affliction.[20]

The inquest jury deliberated and found that a "shot" placed to break coal away from the coalface "blew out" and expended its explosive force outward into the shaft instead of being dissipated in breaking coal. It seemed that the original explosion had also touched off surplus powder lying nearby which intensified the force of the blast. No blame, they said, attached to the management of the mine.

They recommended, weakly, as their knowledge of the situation was so imprecise, that blasting should be overseen by competent men and less powder should be used. The implication was clearly that the accident had been caused by the white miner who had improperly installed an explosive shot. And yet the jury could not leave it at that and added the comment that only white "trappers" (openers and closers of doors) should be employed — how this would have prevented the explosion or lessened its effect was not made clear and may have been a result of the agitation against Chinese labour and the almost instinctive tendency to assign some of the blame to the Chinese if at all possible.[21]

If the agitation for the removal of the Chinese from underground work had been fierce after the Nanaimo disaster, the blame for which had also been assigned to white miners, it now became white hot. The miners firmly believed that the Chinese were at the root of the safety problem and they must go. But if the Chinese were really the problem perhaps a part of the remedy lay in the miners' own hands for they were employers of Chinese on a scale which might have been beyond that of the colliery owners. In the Wellington accident Dick's final report listed twenty-nine dead Chinese of whom eleven were listed without comment but the names of the remaining 18 were linked to the names of dead miners by such notations as: David Gordon's Chinaman No.72 Ah Len and John Stewart's Chinaman No.64 Fong.[22]

During the rest of 1888 there were many more accidents on a smaller scale than the Wellington Disaster: eight were caused by coal cars in the mine and two of these were fatal; two deaths resulted from falls of rock, six men were trapped by falls of coal, men were injured by faulty shots, by accidents in the shaft, and one man was killed by cars on the railway. No Chinese worker was involved in any of these accidents and there is nothing to show that they played any part as cause.[23] In an enquiry later held on Chinese immigration Francis Deans Little, General Manager of the Wellington Colliery, would say in regard to the safety of the Chinese as underground workers that for some years his company operated a mine with three white safety workers and one hundred and fifty Chinese and they had not had a single fatality.[24] The frequently repeated comment of the Inspector of Mines in his annual reports with regard to the cause of accidents in the mines of the area was that he could not discover that any blame or negligence could be attached to anyone indicates that he did not see the Chinese as a cause greater than that of other employees. His other often-made comment was that, although there was provision in the Act for teams of miners to inspect the mines for safety, this was only done by the miners working for the Vancouver Coal Company, while the opportunity to do so was consistently ignored by the men of Wellington. It was his opinion, accurate or not, that the owner of the mines would welcome such miners' inspections.

The white miners were not to be convinced by evidence contrary to their own opinions and took their own action. The inquest finding was flatly rejected, partly on a claim that the evidence of the miners' committee had not been taken in full, and at a mass meeting of the Nanaimo area miners a motion was carried that they would refuse to work in the mines until the underground Chinese were removed.[25] The next morning a delegation met John Bryden and were told that "they could resume work at once without Chinese." Robins offered the same for the Vancouver Coal Company and at an evening meeting, amid applause, the men decided to return to work[26] — but that was only at Wellington and Nanaimo, the Chinese "problem" was far from being at an end. While the miners of Wellington and Nanaimo, spurred on by a second disaster, were taking their own action against the Chinese and clearing them out from underground by agreement with the colliery managements, the Provincial Legislature was still struggling with petitions from the Nanaimo Disaster of the previous year. A Select Committee was formed and heard a great deal of testimony from miners regarding specific acts of a dangerous nature committed by Chinese, which apparently made little impression for when it was later proposed to amend the Coal Mine Regulation Act, by adding the words "and no Chinese shall be allowed to work underground in any mine to which this act applies," the amendment was soundly defeated. Among those voting to retain the Chinese as underground workers was the Member for Nanaimo, Robert Dunsmuir, who under pressure had been obliged to remove them from his Wellington mine.[27]

The public debate on the matter showed that others besides the voting members of the Provincial Parliament had doubts about the Chinese as a threat to mine safety. A lover of "fairplay with an earnest regard for the safety of his fellow beings" was moved by the decided bias in the agitation against the Chinese to give what he claimed to be a dispassionate and impartial review of the facts. In a letter, signed only "Thanet," he used the reports of the Minister of Mines to show that in thirty-nine relevant accidents (those involving gas explosions and fires) occurring in the past ten years only three could be blamed on Chinese workers. His appeal to the miners was to get the Chinese out of the mines if they wished, but not by claiming that they were a danger by being there, for if the Chinese, responsible for only eight per cent of the accidents, were dangerous, what was to be done about the white miners who were responsible for the other ninety two per cent? "Thanet" would like to see all the Chinese banned from the province but doubted the constitution permitted their removal by legislation — or their removal from the mines either.[28] Others worried not at all about the evidence, real or imagined, but took a prosaic business view: labour was in short supply and the Chinese willingly did work white labour rejected; it would be an act of sentimental folly to push out the willing workers when replacements could not be found.[29]

The view that the Chinese were not the danger that they were made out to be by the miners seems strong. In the ten years after the Wellington Disaster eighty-one men died in British Columbia mines to raise 9 million tons of coal — a life for every 100,000 tons. For that same quantity a man was slightly injured. Three hundred and four men were said to have been seriously injured — a serious injury for each 30,000 tons of coal. Accidents when they occurred were serious rather than slight. Two thirds of those who died did so from falls of coal and rock; categories which also brought about most of the serious injuries. There continued to be gas explosions but in only three of the ten years were there fatalities from this cause. Mine cars moving through the black tunnels of the mine were an ever present danger and a productive source of injuries. Even the mules used to handle the cars in some parts of the mines added to the accident toll. When they went beserk they could be relied upon to produce serious injuries. Accidents caused by explosions of powder and badly placed shots, the official causes of the two major disasters, were numerically insignificant. There was no apparent pattern to the accidents: in 1889 when coal production was just over half a million tons there were 4 deaths and a total of 37 accidents which was exactly the same number that occurred five years later when the total output of coal was over one million tons.

In 1891, the first year in which production topped the one million ton mark, there were 86 accidents which included 15 miners killed.[30] There seemed to be little rhyme or reason to the kinds of accidents taking place; every category had blanks from year to year and each seemed to have years with unexplainable peaks. The Inspector of Mines' assessment that there was no one clearly to blame seems reasonably justified. The men had been blamed for carelessness, the Chinese had been blamed because they couldn't speak English — though neither could a great many immigrants who had the good fortune to have a white face — and the owners and managers had been blamed for parsimony and callousness. Without a doubt many accidents could be traced back to any of the groups but the main cause seems to have been that the miners worked in a dangerous occupation which no safety measures could ever make safe.

But there was blood on the coal, and it was the miners's blood. They were concerned and had good reason to be so, in spite of the "factual" meanderings of outsiders. Rightly or wrongly, they believed the Chinese to be responsible for many of the dangers of the mines and on that belief their actions were based.

The Vancouver Coal Company had removed all its Chinese workers from underground, but the same could not be said of the mines controlled by Dunsmuir which were to be found not only in Wellington but also in Extension and South Wellington and increasingly in the newly developed coal deposits of the Comox Valley. The workers in these mines had no protection from the Chinese safety threat and set about getting a reconsideration of the defeated amendment which would have barred Chinese workers from underground employment. A petition, signed by over three thousand miners and residents of Nanaimo, Wellington and Comox, asked the Legislature of the province "to cause an effective legal provision to be made to exclude Chinese and Japanese from working underground in any of the collieries of the province." The petition pointed out that Wellington and Nanaimo had been almost free of accidents since the Chinese had been barred there. As some indication of the growing strength and unity of the labour movement, a petition with almost identical wording was submitted at the same time by the Vancouver Trades and Labour Council.[31] Significantly, although the safety complaints of the past had always been directed against the Chinese, the newly arrived Japanese were now included as undesirables. They, it seems, had been included not because they had been proven to be unsafe, but simply because they were Oriental and so tarred with the same brush as the Chinese. Their inclusion seemed a small point, but it was the one which was going to keep Orientals working underground for many years to come.

7. *John Chinaman*

The Chinese "problem" that would later plague the west coast of Canada was a very small part of a much greater whole. European territorial and trade expansion in the nineteenth century played an important role in increasing racial contacts and helped to stimulate population movements from the crowded old lands of the world to the almost empty new. Around the Pacific Rim the new and empty lands were open to immigration and by the middle of the century there was a growing flood of Chinese into California, spurred first by the discovery of gold and then amplified by the labour demands of railroad construction.

In 1848 there were fifty-four men of Chinese descent in California — within three years the number had grown to ten thousand. At first government officials welcomed them, but the welcome soon faded. The California State Governor of 1852 referred to the Chinese immigrants as a unique and desirable addition to the community that would greatly help in developing the resources of the land. Four years later another Governor, after a period of sustained immigration, was speaking of coolies who were ignorant, avaricious and dangerous. In 1862 the Governor in office wished to discourage settlement of the "inferior race" as they would have a bad influence on the superior.[1]

Other branches of society saw the "problem," if it was such, in different ways. There was no doubt in the minds of many business leaders that the Chinese were highly desirable immigrants. Work was to be done, white labour was scarce and expensive and, what was even worse, stridently vocal about poor working conditions. Profit was to be made, and there was no better way of making it than by employing the docile Chinese for long hours at low wages and with little need to consider working conditions. Open the gates wide and let them all in was their attitude. White workers saw the rise in the number of Chinese as a threat to their own security and in many industries they were proven correct as Chinese became the preferred workers and the whites were pushed out or had to accept lower pay and worse working conditions. Faced with unemployment, depressed wages and the common practice of using the Chinese as strike-breakers the labour unions and "anti-coolie" organizations began to agitate for the exclusion of the Chinese as immigrants. Competition for work led, perhaps understandably, to open violence; for many whites simply being Chinese was sufficient reason to justify ill-treatment. An eye-witness of the first meeting between white goldminers and Chinese workers being shipped in has it that:

> There in the middle, looking mighty skeered, was a lot of fellows with long tails on their heads and short trousers, besides great broad-brimmed wooden hats and a pole like a neck yoke over each shoulder. They were queer lookin' chicks and no mistake. The chaps made a big ring, and yelled and hooted, and quickly began to throw things at 'em to get 'em mad . . .[2]

Bancroft, the historian of the west coast, wrote sadly of the incident, "these freaks of miners started our attitudes towards the Chinese which led to such unhappy results in both the United States and British Columbia, for had no steps been taken in one case they would not have been taken in the other."[3] There was an undoubted link, for when the Chinese immigrant flood flowed north into British Columbia the attitudes of the various sectors of colonial society, the changes of attitude through time, the methods by which it was hoped to control the Chinese problem and even the sequence of events would mirror the American pattern to a remarkable degree.

On the last day of June, 1858 the first Chinese landed in Victoria, the first of many. If they had not found riches in the goldfields of California then, like many thousands of others, they would try again on the banks of the Fraser River. Within a few weeks several hundred

Chinese were scrabbling for gold in the lower canyon, and most of them had first passed through Victoria to obtain mining licences and, to the huge delight of the local merchants, to purchase mining equipment. By January of the next year immigrants were not only coming in from the Californian goldfields but were being shipped out directly from Hong Kong. The owners of the *Victoria Gazette*, alarmed by the sudden influx, were warning their readers of the harm done in Australia and California by this same "peculiar class of newcomers."[4] To the residents of Victoria the odd sight of pig-tailed Chinese, clad in national costume, trotting along with their belongings dangling from long poles carried fore and aft across one shoulder probably gave more cause for amusement than for fear. Not many people were impressed by the *Gazette*'s warning. A newspaper comment published in the following year looked to a bright economic future and, rather fulsomely, described the Chinese immigrants as "pioneers of a wealth yet to be discovered."[5]

Community leaders acknowledged the usefulness of the Chinese while no more than tolerating their presence, which was, hopefully, temporary. Governor James Douglas considered them "not a desirable class of people, as a permanent population, but . . . for the present useful as labourers, and as consumers . . . "[6] Amor de Cosmos, lover of the Universe, but not necessarily the Chinese part of it, wrote:

> They may be inferior to Europeans and Americans in energy and ability; hostile to us in race, language and habits and may remain among us as a Pariah race: still they are patient and easily governed and invariably industrious and their presence at this juncture would benefit trade everywhere in the two colonies . . . when the time arrives that we can dispose of them, we will heartily second a check to that immigration . . . [7]

The ordinary citizens of Victoria had no cause to fear the alien race while the town was only being used as a way station enroute to a mainland destination; but they had their doubts just the same. The Europeans and Americans had found the Chinese to be cruel, ignorant, deceitful and cunning; their uncleanliness and overcrowding a potential source of epidemics. Were they not addicted to opium and gambling? Were not their few women known to be prostitutes, depraved and syphillitic? Were not they common thieves who regarded truth as a sign of weakness? Why should the people of the west coast colony, viewing the Chinese reality through the distorting filter of a ready-made stereotype, find them to be otherwise? What they expected to find they would eventually find.

Politicians worried about the creation of a solid and impervious alien community in their midst. Immigrants were supposed to make the community their permanent home, to adapt, to blend in and to participate. Clearly, the Chinese were not immigrants of that kind. They brought no wives, no children, they put down no roots, they remained apart. They had little choice. Physical assimilation was unthinkable to the white population, and so the Chinese seemed destined to remain isolated, unchanged and unchanging, visitors from a distant land to which they fervently hoped to make an early return. The politicians might have worried, but, lacking pressure from the voters, took no decisive action against the Chinese in the early years of their presence. When they finally did decide to move, it was too late, for by then British Columbia had become the province extending the boundaries of Canada from sea to sea. Whatever the local legislature considered appropriate action against the Chinese was subject to approval by the Dominion Parliament.

From the first day of their entry into British Columbia the Chinese were victims of racial discrimination; however during the first years of their presence it was on an individual and sporadic basis. Proposals for discriminatory legislation arose from time to time, but always there were those in the community who strongly opposed such action. The first suggestion that there should be official discrimination came from the white miners of the Fraser River who, paying a tax to have their animals travel over the rough roads to the goldfields of the interior, suggested that Governor Douglas could better raise money for road building by imposing a head tax on Chinese immigrants. So intense was the opposition to the "mule tax" that the Governor wisely dropped it. He avoided taxing the Chinese by raising his road building money in other ways. In 1860 the question of a head tax was raised at a meeting held in Victoria when Hugh Shanks, a carpenter, declaring the Chinese to be a "nuisance, a curse and a moral scourge," proposed that such a tax be levied upon them. The proposal, greeted by a mixture of "hisses, applause and cries of no, no," passed; but not until an amendment which asked that the word "Chinese" be dropped from the motion had been defeated. Local architect John Wright told the meeting that he believed such a tax could not be collected from Chinese immigrants any more than it could be collected from any foreigner.[8] The proposal had no wide support outside the meeting hall and achieved nothing. The *British Colonist* was stirred to express its support for the free immigration of Chinese on the grounds that they purchased large amounts of clothing, boots, shoes and mining implements and so were of value to the economy.[9] Matthew McFie, a writer familiar with the early days of Chinese immigration, also saw some economic advantage in the Chinese presence and did not believe that they would become a problem. Although he had no very high opinion of their cleanliness and had no doubt that they were vastly inferior to whites, he did find redeeming features. The "tendency to look coldly upon the immigration of Celestials" because of their supposed effect on the price of white labour he found to be unjustified. After all, they were capable only of performing menial tasks which white men preferred not to do and so there would be no direct competition for jobs. He saw the Chinese as industrious and law-abiding and doubted that they would ever arrive in the country in numbers greater than could be absorbed for they well understood the law of supply and demand. True, they preferred the manufactures, clothes and foods of their own country; still they had to buy some things locally and this helped the economy. As to the complaint that they sent their money out of the country as fast as it was earned, the profits accumulated by their employers stayed in the country and it was money that would never have been amassed in the first place if Chinese labour had not been available. McFie suggested that the Chinese be treated with gentle forebearance for they were a less favoured part of the human race and barring them from exercising their civil rights could only bring dishonour to the community.[10]

As goldminers, often existing on the abandoned claims of white miners; as servants working in affluent households; as laundrymen to the community and as employees engaged in menial tasks the Chinese had given no reason for most of the population to become alarmed. There existed a vague and amused tolerance with only the rabid extremists complaining about what might happen if the Chinese moved in any way to give direct labour competition. In the early days simple racial intolerance and stereotyped perceptions had brought about a dislike of the Chinese as people of an inferior status; it would take direct economic competition, or the perception of it, to give that dislike point.

For the coalminers of Nanaimo direct economic competition with the Chinese became a distinct possibility in 1867 when the mine owners brought in a few to work on jobs above ground. Victorian mine owners saw labour as a commodity to be bought and if Chinese labour could be obtained at less than half the cost of white labour then to employ them made good sense. The miners did not share the enthusiasm of the owners, and, although it seemed that they were willing to accept a relatively small number of Chinese above ground, they certainly had no intention of allowing them down the pit shaft. With a grim and heavy-handed attempt at humour, the colliers threatened with violence "the first Chinese who forgets his Celestial origin so far as to descend to the "bottomless pit" of a coalmine . . ."[11] In the same edition of the *Colonist* the editor went rather overboard in his comment and expressed the hope that, now that the Chinese had moved in, it would be possible to make some arrangement to keep the white population of Nanaimo. The comment was far too strong, but the fact remained that Chinese economic competition was starting to be a reality.

The decade of the 1870s brought frustration after frustration to those politicians who wished to control the entry and employment of Chinese immigrants. Their attempts were many but their successes limited. Early in 1871 Arthur Bunster, brewer and politician, rose in the legislature to propose a $50 poll tax on all Chinese, solemnly assuring the "house" that he "had been informing himself on the Chinese question for the past two or three days." His claim that there was wide-spread support for such a tax was greeted with open disbelief. One member speaking against the proposal saw no sense in barring any class when the population was so still so very small. Another opponent "knew a Chinese boy in New Westminster School who was as bright as any white boy" and he roundly condemned Bunster's proposal as "unjust, impolitic and impossible." The Attorney General assured the assembly that it would be useless to send such a bill to the Governor for he would not sanction the special taxation of any class. As defeat was inevitable, Bunster withdrew his motion.[12] Bunster's fanaticism about the Chinese was too early and earned him only ridicule. Two days after his motion failed, a letter headed "How Muchee, John?" and signed "Mongolee" was printed in the *Colonist*. The writer, referring to Bunster as the "festive Arthur," suggested that in his attempts to either "distinguish or extinguish himself in the Council" he might well have proposed a tax of $500 rather than $50, for one would have been no more constitutional or collectable than the other. "Mongolee" concluded on an acid note by asking how much per head should Bunsters be taxed if Chinamen should be taxed $50?

But the day when almost all west coast politicians would have to reflect the concerns of their constituents by an almost automatic opposition to the presence of Chinese workers was coming. American workers' dismay over the use of Chinese immigrants as strike-breakers was soon mirrored by the mine workers of Nanaimo. In May, 1871 the Vancouver Coal Mining Company brought in another sixty-two Chinese workers, partly for financial reasons but also, as Samuel Robins a manager of the company later admitted, because "my company had been suffering from a strike of white labourers, and we accepted the Chinese as a weapon with which to settle the dispute."[13] It was reported, rather remarkably, that the arrival of the Chinese had caused no ill-feeling even though it was "naturally distasteful" to the white miners.[14] No ill-feeling except on the part of the newspaper's editor who, given time to observe and reflect on the Chinese move to Nanaimo, wrote a slashing attack on those he referred to as "long-tailed, almond eyed gentlemen," "Heathen Chinees" and "rice eating aliens." The Chinese, of whom he assumed two equalled a single Caucasian in work value, now numbered one hundred at the Nanaimo Pit and must, therefore, have deprived fifty white workers of employment and their families of food. He claimed that, although they had to buy some things locally, their value to the economy was not more than that of "ten Englishmen." Readers were invited to multiply the effect of the Chinese mine workers in Nanaimo by assuming the same situation in every community in the province and they would then see the danger of the dreaded "yellow tide" which would engulf the land bringing all to its own level.[15] At first glance the idea of a "yellow tide" seems ludicrous as the census of the previous year had counted only fifteen hundred Chinese in British Columbia out of a total population of thirty-six thousand. The bare figures give a misleading impression. Of the thirty-six thousand people, twenty-five thousand were Indians who played an insignificant role in the economic life of the province. The removal of the Indians as an economic factor left the fifteen hundred Chinese to be considered in relation to nine thousand whites[16] but the Chinese were almost exclusively male workers, while the white population contained both sexes and all ages. In Nanaimo, a village of about eight hundred people in total, the one hundred Chinese must, indeed, have seemed like a flood. An endless flood at that, for by 1880 in Dunsmuir's Wellington Colliery alone 266 Chinese workers were employed; seven more than there were whites.[17] But that was still in the future.

In the first session of the Legislature held after the region had become a province of Canada, John Robson stepped forward to repeat Bunster's $50 poll tax proposal and had it soundly defeated. Undaunted he rose again twelve days later to propose that the Chinese be barred from employment on provincial and federal works. This too was defeated. If the defeats had any significance, it was in that they put off for a while the unpleasant possibility that proposals passed in the provincial legislature might well be disallowed by the Federal Parliament. Two years later Bunster and Robson returned to the attack: Robson again proposed a head tax and Bunster, not wishing to discriminate against the Chinese, proposed a tax on those residents wearing their hair in a queue or pigtail. The time was still not yet ripe and both proposals were defeated, in part because they were judged to be unconstitutional. The threat of disallowance was becoming an important factor and one that had the effect of protecting the Chinese against discriminatory legislation.

In local politics the Chinese were beginning to have an effect. In Victoria, to the dismay of many inhabitants, it seemed that a mayoralty candidate had been elected mainly by a block of Chinese votes. Seeing this, the people of Nanaimo decided that they would not be dominated in the same way. When the next local election came round, the Chinese, although entitled to vote, were simply not allowed near the ballot box. This action drew a commendation from the editor of the *Colonist* who applauded the good sense of the men of Nanaimo who had managed to avoid the degradation of allowing "heathen slaves" to have an equal voice in the management of the town's affairs.[18]

The coal miners of Nanaimo were not alone in their complaints about Chinese immigrants. There was a brief flurry of excitement when it seemed that the importation of Chinese workers would bring direct competition to the white miners of the Cassiar region. At a meeting held in Victoria to protest the coming of more Chinese a speaker, no doubt with a healthy respect for his own assumed wit, regretted that they did not indulge in wine, brandy, whiskey, beer and cigars, for this meant they would not contribute their fair share of taxation. Even worse than that potential loss of taxation revenue would be an even greater loss, in that the flood of Chinese would cause the rest of the population to lose their vices and vanities. And what, he asked, would be the use of free schools if white men were brought down to the level of the Chinese, for rice cooking would be all that it would be necessary to teach. Robert Smith, a member of the provincial legislature tagged the Chinese as the "grass-hopper of British Columbia" and put forward as a certainty that they would eventually "eat the white population out of house and home." He had read of a boatload of Chinese women landing in California; the thought of such an event in British Columbia filled him with dread. A petition was prepared asking the Legislature to impose a head tax of $30 on each Chinaman entering the Cassiar diggings and 380 men added their signatures.[19] The fear of the "yellow flood" was well to the fore; it was obvious that public opposition to the Chinese was rising fast and the demand for legislation against them growing.

The matter of controlling the entry of Chinese into Canada dragged on interminably. The main reason for delay was that from distant Ottawa the Dominion Parliament saw the Chinese problem as one very remote from the affairs of most of the inhabitants of the country. Of the Chinese in Canada only a handful were to be found outside the boundaries of British Columbia. It was not a matter that could be expected to receive much attention from politicians, the vast majority of whom had never set eyes on either British Columbia or a Chinaman. And British Columbia, it seemed, was pulling in opposite directions at the same time: there had been threats of secession because the railroad had not been built and yet the people of the province were advocating cutting off the supply of cheap labour necessary for its completion. The pressure was building for action — or, as the Prime

Minister saw it, the time was ripe for delay. To obtain the withdrawal of a motion seeking to prohibit Chinese entry into British Columbia Sir John A. Macdonald set up a "Royal Commission to Inquire into Chinese Immigration into British Columbia."

Joseph Adolphe Chapleau, Secretary of State of the Dominion Government, and John Hamilton Gray, Supreme Court Justice and a resident of British Columbia were appointed as Commissioners. At the first public meeting of the commission, Dr. Chapleau stated that he wished to have information on the advisability of prohibiting or restricting Chinese immigration.[20] Judge Gray, rather more forthrightly, said the aim of the commission was "to obtain proof that the principle of restricting immigration is proper . . . " Such proof was necessary for the guidance of the many members of the Dominion Parliament who had no knowledge of the problem but would need to justify whatever action they took to their constituents.[21] The constituents, presumably, knew less about the problems of the west coast than did their parliamentary representatives.

In spite of assurances that they would find a solid and implacable opposition to the presence of the Chinese, the Commissioners later reported that three main opinion groups existed: a vocal and strongly prejudiced minority who would only accept the complete exclusion of the Chinese; what they described as an "intelligent minority" who believed that the law of supply and demand would take care of the problem; and a large majority concerned mainly with order, sanitation and health, who favoured moderate restriction.[22] Evidence taken in the Vancouver Island coal mining district[23] showed a not unexpected division in that worker opinion was for exclusion, while owners and managers, no doubt a part of the "intelligent minority," opted for the economic solution free of all restrictive legislation. Even so, on both sides, there were exceptions which fell between the extremes.

Robert Dunsmuir M.P.P., proprietor of the Wellington coal mines, gave his evidence as a resident of the province for thirty-two years and as the employer of seven or eight hundred men, white and Chinese. The whites, he explained, were employed as miners and in skilled occupations, while the Chinese did manual work. He found the Chinese to be hardy and industrious, temperate and peaceable, frugal and saving — except when they gambled. They kept their contracts and fulfilled their engagements. He had no doubt that in the early days, when white labour was expensive and scarce, the Chinese had filled a great need and had made a material contribution to the development of the province. They were still needed, in his opinion, because they did work refused by others and at wages that made mining pay. If they were removed, there would be a serious effect on the coal industry which would not then be able to meet the competition in the principal market of San Francisco. The Chinese had not slowed white immigration as far as he could see, they did not cause suffering by their economic competition and had no ill effects on the morals of respectable white mine workers. There should be no legislation regarding the entry of the Chinese into Canada; they would not come in numbers greater than the labour market demanded. Nothing would convince him that the Chinese were any kind of a drawback to the province and the agitation against them was "largely political." As might be expected the Dunsmuir managers followed strictly in the footsteps of the master in leaving the door open for Chinese immigrants. Robert Scott would offer no opinion as to how many should be allowed into the province and remarked only that the Wellington collieries needed four hundred. John Bryden thought the number admitted should depend only upon the demand for labour. They had little to say of the racial characteristics of the Chinese whom they found to be industrious, sober and law-abiding; "more so than the same class of white labourer," said Bryden. It often seemed that if one coal company saw black, the other saw white, and it was no different when they gave evidence on the use of Chinese labour in the mines. Samuel Robins of the Vancouver Coal Company had no great use for Chinese workers. In direct contradiction of the Dunsmuir attitude he stated that he saw no reason why they should be retained, adding only the caution that there would be some inconvenience if they were suddenly removed. His suggestion was to limit Chinese immigration by the imposition of a $50 poll tax. He feared that a continuation of the current situation would lead to open violence and added that "we have found it difficult to protect the Chinese from ill-usage."

Among the respondents to the twenty-seven questions posed by the commission were J. Pawson, John Metcalfe and James Young who, although not identified by occupation, clearly expressed the miners' view of the Chinese. They believed that there had never been a need for Chinese workers and only "men having no object beyond their own self-aggrandizement and selfish greed'' had welcomed them. Even now the only people in favour of retaining the Chinese were the monopolists who desired them as a standing threat to white labour. The opposition to Chinese immigrants was of long standing; it had been started by goldminers when the Chinese first appeared and had been carried on ever since by white working men as well as by merchants, intelligent men of the middle classes and by the public. As far as Pawson and Metcalfe were concerned the Chinese had made no contribution to the development of the province. James Young, who had also signed the Knights of Labor brief to the commission, was a canny enough union man to say that the Chinese must have had some effect for all labour contributes to the development of any country. In spite of what the managers had to say they, the white workers, knew only too well what effect the Chinese had on job opportunities. Many men were out of work and others could only find work of an irregular and uncertain nature. Boys leaving school were unable to enter the mines because jobs normally done by boys had been given to the Chinese; even those Orientals employed as household servants had cut off domestic employment for the daughters of the working classes. Not only had the Chinese presence discouraged immigration; it had also caused a loss of population because, James Young claimed, there had been ''hundreds compelled to leave the province from the impossibility of obtaining work.'' The workers were not at all shy about repeating the litany of complaints about the assumed racial characteristics of the Chinese which had become so familiar over the years. They lived in wretched hovels with such a neglect of sanitation that they were a menace to public health; they were the greater part of the criminal element of the province; they were immoral and takers of opium; what women they had were prostitutes, and so on. Again James Young had an original point of view in that he blamed much of the problem on the lack of women which caused Chinese men to become depraved of intellect with a standard of morality immeasurably below that of the white population. But he did not recommend the admission of Chinese women to cure the problem — and neither did anyone else.

Pawson and Metcalfe had a simple solution to the problem: the Chinese should all be forced to leave the province and no more allowed in, the only cost would be that ''the the monopolists would not get so rapidly and unwieldily rich." Young once again took a more moderate view and said it would be to the good of British Columbia if the Chinese could be induced to leave, but he would not "strain the laws or deal unjustly" with them to achieve that effect. Touching on the real crux of the matter, he noted that "people in their haste to grow rich, care not how" and would employ cheap labour unless prevented from doing so by carefully framed laws.

The Nanaimo Knights of Labor struck a patriotic note with their expressed desire to see a united and powerful Canada in which British Columbia, so well situated for trade and with vast natural resources, would grow rich and populous. It only required the right sort of immigrants: intelligent and independent people "akin in blood, who speak a common language" and who have similar social characteristics and hold democratic political beliefs. The Chinese, "a great cancer on the social, industrial and political system" must be checked. Undesirable and dangerous, lacking in understanding of popular government, separate and distinct from the majority of the population, making little financial contribution to the community and eventually draining off wealth on their return to distant China, they had no part to play in the creation of a Canada strong, free and prosperous. Having appealed to patriotism the Knights of Labor then turned to "other and higher reasons," which seemed to be inextricably bound up with their own economic self-interest. The competition of the Chinese had, they claimed, driven wages to the point of degradation for white workers and their servility had only encouraged the oppression of the tyrannical employers who would be delighted to extend the practices they used toward the Chinese to their much less compliant white employees. The situation was bad and becoming worse; between 1881 and 1884 the white population increased by only fifty per cent, while the Chinese population grew by over two hundred per cent. The time was not far distant, they said, when the bulk of the mining population would be Chinese; already they outnumbered the whites at the Wellington Colliery and white boys were denied jobs at the mines for those suitable had been taken by "the ubiquitous heathen." They saw a dismal future: "Soon there will be no resource left to us but to fly from evils we cannot combat" or we shall be "compelled to take up arms to preserve ourselves and our liberties." They were not begging, but demanding as a right that the further immigration of Chinese be at once stopped in the name of justice. Without naming him directly, the Knights of Labor attacked Robert Dunsmuir, a man of "princely fortune" whose all-pervasive influence reached into every part of the provincial government so that its existence was threatened. A man who ensured that white workers must constantly face the economic competition of the Chinese, while at the same time effectively preventing others from emulating his rise to wealth through his control of the coal-bearing lands. In denying others the right to open coal mines he had prevented the expansion of Nanaimo into a great manufacturing city and so it remained "a languishing village . . . her sparse population largely composed of Chinese." The labouring classes were left quite unprotected against the "killing competition of a degraded race" while a few individuals, already too rich, were in possession of the natural resources of the country and so placed beyond the reach of all competition.

The Nanaimo Trades Association, no less patriotic, pointed to the efforts of the Dominion Government to infuse a national sentiment by a protective policy of "Canada for the Canadians." As "nationalists, as protectionists, as British Columbians" the members of the Association appealed to the Dominion Government to give them protection from unfair foreign competition, or, this no doubt said with tongue in cheek, to allow the Chinese to have large tracts of coal-bearing lands so that the capitalists might get a taste of Chinese competition for themselves. The coal lands monopoly, the chief shareholder of which (Robert Dunsmuir) had made a "visible fortune of $2,000,000," had killed the chances of Nanaimo becoming an industrial city. Had competition been allowed the price of coal would have been cheaper and manufacturing "rendered not only feasible, but in a manner compulsory."

In fact, while making it clear that they did not like the Chinese and wished to be rid of them, the labour organizations said little about the racial problem. Their main concern was with the coal-owners, Robert Dunsmuir in particular, and in this they were quite right, for the battle was between capital and labour and not between labouring groups. Had they, for instance, insisted that regulations be made setting a common pay scale for all mine workers, regardless of race, then the white workers who claimed such superiority over the Chinese would have had the advantage and the problem would have solved itself in their favour. As it was they had recognized the reality but would continue to fight the symbol.

Early in 1885 the commissioners made their separate reports which were generally in agreement one with the other. In their desire to ensure a bright future for the province the commissioners aimed to protect existing industries and so favoured a policy of selective admission rather than erecting legislative barriers that would cut off labour supplies. They had, in fact, come down on the side of the industrialists and entrepreneurs — the Dunsmuirs of the world. To placate the moderate majority, they would bar from entry those Chinese who were found to be diseased, deformed or impoverished, while at the same time forming a commission which would register the Chinese already in the province and then see that the laws were enforced among them and the taxes due collected. To those who favoured exclusion or a stiff poll tax as a discouragement to Chinese immigrants they offered the almost insulting sop of a poll tax of $10.

The talk the commissioners had heard of open animosity toward the Chinese grossly exaggerated the situation. What animosity there was seemed to them to be centred around the collieries where the competition to white labour was most concentrated, and even there abuse was verbal rather than physical. This they took to be an indication that the level of Chinese competition was not intolerable to the white mine workers. So few were reports of violence against the Chinese around Nanaimo that the conclusion of the commissioners might well have been valid. When there were incidents they were usually minor in nature and of the kind carried out by thoughtless adolescents.

The recommendations of the commissioners were seen as being weak and useless, and the more anti-Oriental sections of the public were not slow to say so. In Victoria a crowd estimated at well over one thousand met in protest carrying banners with such messages as "No Yellow Slaves Shall Eat Our Children's Bread" and "Cut out the Chinese Cancer." As far as the protesters were concerned the commission had not fulfilled its task, which was simply to get rid of the Chinese. By the end of May an "Anti-Chinese Union," along the lines of the "Anti-Coolie" organizations which had earlier sprung up in the United States, had been formed; ironically at a meeting held in Harmony Hall.[24] The provincial government was also displeased and again passed its own law to discourage Chinese entry only to have it disallowed as usual. But the federal government, now almost free of the necessity to get cheap labour for railroad construction, moved to impose a $50 arrival tax on all Chinese immigrants as well as limiting the number to be carried on each ship to one for each fifty tons of the ship's tonnage.

The rate of Chinese entry into British Columbia slowed due to the new immigration rules, but at the same time the employers who did import Chinese workers were given an extra hold over them, for it was most likely they who advanced the $50 entry tax — for future repayment and until the debt was cleared a guarantee that the debtors would do exactly as ordered. In any case the slower rate of entry did not help the mine workers of Nanaimo very much; their problem was the Chinese workmen already in the mines. At a meeting said to have arisen through the "spontaneous action of the whole community" a committee was appointed to collect money and use it to print cards for display by shopkeepers and householders to show their anti-Chinese feelings.[25] This was only a ripple of displeasure which would swell into a great tide after the Nanaimo and Wellington mine disasters for which the Chinese underground workers would be blamed — even though the inquest juries in each case would point to the incompetence of white miners. The accidents triggered a decade of action and counter-action, legislation and disallowance, a decade of sound and fury none of which brought a solution of the racial problem any nearer. The miners' approach, through their legislative supporters, was to bring about amendments to the Coal Mine Regulation Act which would bar the Chinese from underground employment. Such an amendment was passed after miners petitions had been received by the legislature; however, in the professed belief that the amendment would be declared unconstitutional, the Dunsmuir's simply ignored it and continued to use Chinese workers underground. Further petitions from the miners included the newly arrived Japanese as persons who should also be banned from underground work in the mines. In acting to please the miners the politicians stirred up for themselves a hornet's nest of international proportions.

His Japanese Majesty's Imperial Government took exception to the attempts of the Legislature of British Columbia to deny employment to Japanese nationals and the Japanese Consul-General in Vancouver made known his Government's displeasure in a letter asking the Governor-General of Canada to disallow the offensive legislation. In the Japanese view, no satisfactory reason had been given for the flagrant discrimination that was to be practiced and when there were only about one thousand Japanese in the province, less than one-tenth of the number of Chinese, they did not think that a satisfactory reason could be found. With a sharp touch of asperity and, perhaps, rather more than a touch of personal racial bias, the consular official found distasteful the tendency of the Legislature to lump the Japanese in with the Chinese when it was clear that their education and character, customs and manners were entirely different.[26]

Two months later the British Prime Minister wrote from Downing Street to acknowledge the receipt of documents about the Japanese protest which the Governor-General of Canada had sent to him. He wished the Governor-General to make it quite clear to his ministers that the legislation was "repugnant to the people and Government of Japan" and advised that it could reasonably be replaced by something along the lines of an enclosed copy of the "Natal Act" which seemed to have served the purpose of limiting similar immigration into Australia.[27] The Natal Act, apart from the general prohibition of those found to be physically, mentally or morally undesirable, required that each immigrant should write out an application for admission in any European language. The Japanese, in an attempt to cover the possibility that the legislation might not be disallowed by the Canadian Parliament, also sent a letter to the Foreign Secretary of the British government enlisting his aid.[28] Seldom had there been any protest from China about the treatment of Chinese nationals in Canada but the Japanese government was a different matter. The Legislature of British Columbia was teetering on the edge of very deep waters.

A few years later the appointment of another Commission of Enquiry showed quite clearly that the Oriental problem was far from over. The new evidence it collected from the Vancouver Island mining community was a good indicator of how much, or how little, the attitudes of the various concerned groups had changed in the years since the first report appeared in 1885.[29]

Samuel Robins was the only person from the mining community to give evidence before the new Royal Commission who had also given evidence before the previous Commission. His views had changed very little. Following the disasters of the late eighties his company, at the urging of the white miners, had removed the Chinese from underground employment. He had agreed with that decision because he believed they lacked the white miner's courage in adversity and would be panic-stricken and useless in an emergency. Sooner than take the Chinese back underground, his company would suffer ruin. The question of the use of Chinese workers at the surface was not quite so clear cut. The use of oil was increasing and there were some difficulties in marketing coal; if the lowly paid Orientals were not available, it might mean that the white workers would have to accept lower wages. However, he could see no evidence to suggest that the Chinese had changed their living practices over the years — they were still as Chinese as ever — and he thought action should be taken to limit their numbers before the Mongolian flood reached unmanageable proportions.

James Dunsmuir, who by that time had succeeded his father as head of the coal empire, chose not to appear before the commission to say what his views on Chinese immigration were, leaving the Commissioners to record the rather plaintive comment that "they were not favoured with the views of Mr. Dunsmuir, although requested." As was usual, it seemed that, if the Vancouver Coal Company saw white, the Dunsmuir managers would see black; it was no surprise when they expressed views quite the reverse of those of Samuel Robins. They had never had any trouble with the Chinese underground and were not hesitant about saying that they disapproved of the costly experiment carried out by Mr. Dunsmuir in removing the underground Chinese from the Extension Mines. John Matthews, local manager of the Union Mine at Cumberland, some fifty miles up the coast from Nanaimo, had no preference for "Canadians over a Chinaman" as the Chinese were as safe underground as whites — always providing they spoke English. It was well that he thought so for of the 877 men employed at Dunsmuir's Cumberland Mines 455 were Orientals and 412 white, and the whites were outnumbered below as well as above ground. The managers of the Dunsmuir Mines were as opposed as they had ever been to any restriction on immigration and all stated, in various ways, that without cheap oriental labour the wages of the white miners would fall or the mines would have to close. General Manager Little said that he had never had any objection from his men about the use of Chinese, but admitted that the union had made a demand for their withdrawal:

he drew a fine line between his men and the union and insisted that he did not know whether or not his white workers were union members. The managers of the Dunsmuir Mines said little about the racial characteristics of the Chinese, not because they were free of bias so much as because their concerns were economic rather than racial. When questioned closely the General Manager admitted that although the mines could be operated entirely by Chinese with greater economy it would not be desirable — "I would rather have the whites than the other sort." His subordinates agreed.

There was no shortage of white miners to appear before the commission. To some extent what they had to say was a repetition of what their predecessors had said in 1884. However, since in the intervening period there had been two great mine disasters, they had safety sharply focussed as a major concern. There was still not the slightest doubt among them that the Chinese had been responsible for both events. They suggested to the commission that one way of solving the Chinese problem was for the government to legislate a minimum wage — not so much because they wanted the Chinese to be treated fairly but rather that they believed that, if it became expensive for the employers to use the Chinese, they would naturally choose to employ whites. The Chinese would then be driven out by unemployment.

The two union submissions summed up the prevailing worker attitudes. The Nanaimo Miners' Union wished to have immigration of the "ignorant and therefore dangerous class of worker" restricted or prevented. The laws passed by the provincial legislature to keep the Chinese from working underground had repeatedly been struck down, and they had to rely on the uncertain disposition of the mine owners for protection against the dangerous Orientals. They implored the Dominion Government to come to their aid. The Alexandra Miners' Union was more forthright. Apart from the usual run of safety and economic concerns, they found so deplorable the habits and character of the "Mongolians" that it seemed to them that unless there was a speedy expulsion the white men would all leave the province.

Interesting cross-currents flowed from the evidence. The general manager of the Wellington Collieries in commenting on the desirability of employing Chinese mentioned his experience with two hundred Scottish miners, brought in at heavy cost, only to find that in a short time almost all of them had disappeared to other places. James

Cartwright, a Nanaimo miner, illustrating how the Chinese discouraged other workers cited a group of Scots, possibly the same group, who left because they would not work with Chinese. Andrew Brydon, manager of the Extension Mine, who believed that the Chinese underground were no particular threat, was himself referred to by John C. McGregor who in response to a question answered, "When I would ask Mr. Brydon to go into a place he would say, 'Have you been in there?' If I said yes, he would go in, but if I said a Chinaman had been in, he would not go in." It happened that the resolutions and complaints about the Chinese were presented by miners from mines with no underground Chinese workers. How much stronger the force of the resolutions might have been had similar resolutions also come from mines, such as those in the Comox area, where Orientals were still working underground. Perhaps the revelation of general manager Little that of the 450 Orientals employed the company paid only one third, while the rest were in the employ of individual miners, and at low rates of pay, might account for the omission.

Possibly from long habit and because they were speaking of events some little distance in the past, the focus of evidence was on the Chinese. There were Japanese workers in the mines also, but only one miner was astute enough to mention them, saying he regarded "the Japanese question as more acute than the Chinese question." Right or wrong it was undeniable that another ingredient had been added to the racial brew and as the century closed the miners were as far as ever from having the racial question settled to their satisfaction.

8. Unite

Thursday, July 17, 1890 was a strange day in Nanaimo and the strangest thing of all was a long procession made up of several hundred men, women and children winding its way slowly down the road from the direction of the mining town of Wellington. The procession straggled along until the Court House was reached and there came to a halt. Of the throng, one hundred and thirty had been summoned to show cause why they, as renters of Dunsmuir houses, should not give up the tenancy of the houses they occupied. It was a tight squeeze to cram them all into a court room designed for much smaller numbers, but the need for publicity, if nothing else, determined them to try.

The Dunsmuir brothers, unlike their late father, were not prosecuting selectively, but were getting rid of every last striker in one single action. There was no shred of doubt that they were legally entitled to demand the return of their houses; Judge Harrison made it quite clear that the rights of property must be protected. In his opinion, the strikers had no right to be occupying the houses in question and the plaintiffs had every right to demand the return of what was legally theirs. The question of the miners staying in the houses simply did not arise, all that the court must decide was how long should be given to the tenants to vacate. According to circumstances, the delays were set at times varying from five to twenty days. This time there was no trouble with the evictions; as the moving dates came up the miners and their families moved quietly into hastily constructed shacks or tents. Tully Boyce, president of the Miners'

Union, had already sent a message to the people of Victoria asking for the loan of tents for those families unable to provide their own.[1] Between the appearance of the miners in court and the time they moved into their makeshift accommodation, the *Colonist* carried a brief report concerning five loads of hardwood fittings that had been shipped from Chicago for delivery to Craigdarroch Castle.[2] Joan Dunsmuir, the owner of the Wellington Mine, was seeing to it that her castle should be finished. She would not go short during the strike.

Of the numerous Dunsmuir children, James and Alexander were the only boys. Destined to follow their father into the coal business, they would play a significant role in his coal empire and, necessarily, in the affairs of the province. The slovenly, rough and ragged settlement of Nanaimo was full of contrasts and contradictions, and in it the boys were not shielded from life. Often the miners they met were uncouth in behaviour and habits, familiar with hard language and even harder liquor. While the lower orders feared neither God nor man, other whites, of a "better" class, were often narrow-minded and rigidly bound within a mesh of severe and unyielding religious beliefs. Throughout the community the native Indians roamed, sometimes drunk on "tanglefoot," the illegal whiskey sold to them by unscrupulous white men, and "Klootchmen," Indian prostitutes, were always to be seen looking for business. There were variations of colour and mixtures of races in varying degrees. Small the settlement might be, but it probably held within its boundaries as great a slice of life in the raw as could be found in many places of a much greater size. Both boys spent a good deal of time in the Indian encampments and Alexander, was a particular friend of "Coal Tyee," the coal chief who had provided the first samples of coal from Nanaimo to Joseph McKay at Fort Victoria. "Coal Tyee" was involved in one of the frequent clashes between Nanaimo groups, being banned from the church services at the insistence of a group of "Christian" ladies who objected to his smell. The incident led James Dunsmuir to remark that the old Indian was much to be preferred to the preening old hens in the front pews.

By the time the boys were well into their school days Robert Dunsmuir had become mine manager with absolute power to decide who would work and who would not, a fact which no doubt set the family somewhat apart from lesser people. It is quite possible that in any case the Dunsmuir boys would have been different if only because they were influenced by parents of an unusually strong character: Robert Dunsmuir was hard working, business-like and obsessed with coal and his continuing search for the seam which would give him his opportunity; Joan Dunsmuir resolute, determined, devout and with a good common-sense Scottish attitude to the benefits of a sound basic education. But even in the same conditions and with the same parental influence the boys turned out to be very different. James, the elder, was most nearly patterned on his father: practical, steady and reliable, rather shy and not given to using more words than were absolutely necessary to achieve his purpose. Alexander, on the other hand, was independent, disinclined to take advice, ambitious and, while having the Dunsmuir business sense, more inclined to look outward to the broader scene rather than concentrating on what was before him in Nanaimo.

Their training and employment followed the trend indicated by their divergent characters. James became an apprentice machinist in the mine and later went to the Willamette Iron Works in Portland, Oregon to complete his technical education. He worked about the mine and then, in what seemed an unlikely move for a quiet young man from a tiny west coast mining village, crossed the continent and entered the Hamilton Military Academy in Virginia. While there he met and married Laura Searles and they moved to the west coast to live in Wellington. He first took over the shipping side of his father's business, and then became mine manager at Wellington before handing over to John Bryden and moving to Departure Bay to look after the administrative side of the business. Alexander showed little practical flair and had no aptitude for running a coalmine. His early occupation was assisting his father in real estate deals and acting as a liaison man with government contacts in Victoria. After the 1877 Strike he was sent to San Francisco to bring about his father's dream of not only controlling the production of coal but also distributing it in his major market in such a way as to squeeze every cent of profit from each ton. Joan Dunsmuir was not pleased with the move, sensing correctly that Alexander would get into trouble in San Francisco where she believed there were many attractive and scheming women lying in wait. Within a short time of his arrival he had met Josephine Wallace, the wife of a theatre usher, and moved in with the family. Shortly thereafter the husband left and Alexander and Josephine moved into new lodgings together. They were to stay together; the opposition of his mother to any thought of marriage between them became the central fact of Alexander's brief life. The liaison also had a marked effect on his brother's life and on the firm.

Although the sons undoubtedly became of much greater importance in the business as Robert Dunsmuir moved away from direct control, there was no doubt within his lifetime who was boss. On his death in 1889 the running of the business, in all its practical facets, was entirely in the hands of James and Alexander, but they were not entirely free of parental control, for Robert had left to his wife, and to her alone, all his assets. Sitting in Victoria in her castle, Joan Dunsmuir would continue to hold the reins for many years.[3]

What the sons had taken over on the death of their father was trouble of the Dunsmuir kind — strikes and threats of strikes. In the year of his death Robert Dunsmuir had closed the Wellington Mine on the threat of a strike, as he said, to protect his workers whom he would not have been able to employ again if they had carried out their strike threat. On his death the men had returned to work as though to start off with a clean slate with the new boss. It was a short honeymoon. Little more than a year after the death of the elder Dunsmuir the Wellington miners were enmeshed in a strike which had much in common in its major events with the Strike of 1877 — the treatment the miners received from the Dunsmuir brothers was no whit different from what they could have expected from their father — except in the increased efficiency of its execution.

There were problems in the Dunsmuir Mines and in mid-May of 1890 the miners brought them to a head. At a Saturday evening union meeting it was decided that the Wellington men would work their eight hours "from bank to bank"; demand recognition as an organized body, a part of the Miners and Mine Labourers Protective Association; and insist that the employers should allow a committee of their own employees to arbitrate any dispute arising between an individual miner and a mine foreman. A committee was immediately selected to inform the mine manager, John Bryden, of the resolutions and get his written agreement to them. Manager Bryden, even had he wished to agree, had not the power to agree on behalf of the Company and asked for time to place the matter before the Dunsmuir brothers.

His request for time was refused.

At the pithead on the following Monday morning the miners, who were normally in their work places underground by 7 a.m., stated that as it was their intention to spend only eight hours underground they would not enter the cage and go down the pit before seven. When the time to start work arrived and the men were still on the surface, they were told to go home and return at the proper time on the next day. The company to the apparent consternation of the miners had struck back instead of meekly submitting. A *Colonist* correspondent, tracing the events of the strike at a later date, wrote "the action fell like a bombshell on the miners who had not dreamed for a moment that there would be any resistance."[4] Why miners experienced in the ways of the Dunsmuirs should expect a meek acceptance is hard to understand. In any event the strike was now on — at least in the rhetoric of the mine owners; having been sent home and not allowed to work, the miners claimed that they had been locked out. The effect was the same; no coal was being produced for sale by the owners and no wages were being received by the men.

This was a strike with a difference and many of the Wellington miners were very concerned about the way it had come into being. It was usual for each of the union locals to run its own affairs and decide its own actions by a vote of the members of the local, yet at the meeting a few days earlier all union members had been invited to be present and all had voted. There were suspicions that the men from the other locals had outnumbered the men of Wellington, and as a result that the miners of Wellington were being deprived of their livelihood by the votes of men who had nothing to lose. At a subsequent enquiry Tulley Boyce, the miners' leader, was unwilling to swear that there were more men from other union branches at the meeting than from Wellington, but he was also unwilling to swear that it was not so. Some men pointed out that the union had broken its own regulations which gave the owners the option of settling by arbitration or conciliation. Unlikely as it might be that the Dunsmuirs would have accepted either, they had not been given the chance to refuse. It was the opinion of some men that the union had cut their throats by its treatment of Mr. Bryden. Did the strike come about as the result of a miscalculation of the owners reaction to a specific action or was it the result of deliberate union provocation designed to cause a strike? Even at the time opinions were divided, but that there was a divergence of views pointed up the fact that support for the strike would be a problem.

There were miners who were simply anti-union and saw no reason why the Dunsmuirs, of whom there were "no better bosses on the island," should be expected to cave in to the demands of the union, the leader of which was said by one miner to be a "most determined man, and would sell all he had and imperil those who are most dear to him to carry his points." Some miners contended that the eight-hour day, which they had never heard raised before the union meeting that had started the whole miserable affair, was not an issue because two-thirds of the men were on piece-work and were able to work as much or as little as they desired. But union strength had grown since the first faltering union had come into being thirteen years earlier at the beginning of the 1877 Strike, and, as a conversation between a newspaper correspondent and a striking miner showed, if men could not be brought out of the mine by conviction then fear would do just as well:

Question: Why are you on strike?
Answer: D——— if I know. The fellows quit work and I quit with them.
Question: Why?
Answer: I wasn't going to be called a 'scab' or a 'blackleg', and I didn't want to get out with the gang.[5]

From a union tax on men still working in other mines, each man received as strike pay $13 monthly if single and $16 if married with an additional $2.50 for each child in the family. It was still summer and living in tents and shacks was bearable, but soon the fall would bring increasing rains and then would come winter with more rain and cold and even snow. Then it would not be so easy to put a brave face on it. As one man described the situation it was a "G—— d——— hoodlum strike" arranged by a bunch of sand-lot operators who themselves risked nothing and no man could be expected to see his family deprived of a home and food for very long in these circumstances. The men had lacked the common-sense to keep out of the strike, in his opinion, and now the big difficulty facing them was that they didn't know how to get back to work. All it needed was that the company should bring in a few outsiders to start up the mine and there would then be plenty of the old workers ready to flock back and join them. But there would be trouble for the more determined of the men would resist a return to work.[6]

A remarkable feature of the strike was the way in which the clergy moved in on the topic of the eight hour working day, often keeping a delicate balance as they trod between the minefields of their parishioner's divergent views and interests. The Reverend Jenns[7] contended that labour is a blessing for, "If God had intended man to live without work he would have made their clothes to grow upon their backs, and furnished houses to grow out of the ground." Even so, he thought, our short time on earth should not all be spent working. His conclusion, after a lengthy excursion through the sometimes troubled relations between capital and labour in the Kingdom of Solomon, was that he supported the eight-hour day. He went on to spoil his conclusion by adding that in pioneer countries people should expect to work longer hours and that being so it would be more opportune if the reduced working day was first introduced in the older countries. Reverend Jenns brought his remarks to an end by stating that British Columbia was already "the workingman's paradise," a sentiment which was no doubt of great interest to the miners of Wellington. The Reverend Mr. McLeod[8] was in a most difficult position as he stood in the pulpit of St. Andrew's Presbyterian Church bathed in the light filtering through the stained glass windows donated by Mrs. Joan Dunsmuir in memory of her late husband, Robert Dunsmuir, the same Mrs. Dunsmuir of Craigdarroch Castle who held so firmly to the control of Dunsmuir and Sons against which company the strike was in progress. McCleod made a rather tentative start to his sermon, asserting that a Christian minister as a faithful arbitrator must inevitably receive the ill-will of partisans on both sides. He regretted the millions lost by labour fighting capital and by capital out to conquer labour and saw no reason why the capitalist should wish to hold the workingman in subjection, or why the workingman should force his views on the capitalist and exact vengeance if they were not accepted. Contracts were made to regulate industrial relations; however, in his view, they were not made from positions of equality as the worker was more dependent on the capitalist than the reverse, and so public opinion should count in redressing the balance. He refused to doubt that the masters and men would end up shaking hands over their differences. For his part he favoured the eight-hour day, if only because the change from ten hours to nine had created no great problem and so why should a change from nine to eight be any different? Labour lightened was, to him, labour improved. Not until a week after the lesser ecclesiastics had held forth did the Right Reverend Bishop Cridge weigh in with his opinion[9] taking as his text, "Man goeth forth to his work and to his labour until the evening." He failed to define when evening occurred to end the day. He also claimed not to be an arbitrator because a Christian minister had no warrant for settling worldly disputes. As he saw it, the men had not had any of their rights denied but he seemed to be vaguely on their side because the judgement of God always appeared to be on the side of the poor rather than the rich as the rich have the upper hand. He could see no impoverishment of society if the men were to be granted an eight-hour day. He confessed that the idea of dividing the day into three equal portions; one third for labour, one third for improvement and devotion and a third for rest, was captivating. He did not elaborate on the prospects of attaining the other two thirds if the working one third was set. The problem must really be solved by the workmen and their employers he concluded and hoped that they would do this by observing the law of God.

The reverend gentlemen, with their pious hopes for tolerance, understanding, and brotherly love, aptly demonstrated the pure reason of the uninvolved but it is highly unlikely that any single participant on either side of the dispute was influenced by them. It seems that having been approached by both sides to preach on the strike, the clergy had deliberately chosen the path of amiable vagueness well knowing that it was the intention of both sides to derive advantage for themselves. Their qualified support for the eight-hour day was such as to have added no more heat to the situation. In any case the clergymen, and many others, were on the trail of a good red herring: the strike was not about the eight-hour day and it never had been; it was about union recognition. Confirmation of the real cause came from no less an authority than the union secretary who, when in a chance conversation with the Dunsmuir brothers, insisted that the prime condition of settlement was union recognition. Predictably, Alexander Dunsmuir stiffly replied that the union would never be recognized.[10] The cause was later confirmed by the Select Committee appointed by the Provincial Government to examine the reasons for the strike. The committee found that "the primary cause of the strike at Wellington in May last, was the desire of the miners to be recognised as an organization known as the Miners and Mine Labourers Union, and the refusal of the management of the Wellington Collieries to recognise or treat with such an organization."[11] In making such a finding they were only repeating the words of union president, Tully Boyce, who as the first witness before the committee said in answer to the very first question:

> . . . as far as I know, the cause of the strike is that the Company refuses to recognize their workmen as an organized body. The question of hours, if I understand it properly, is entirely out of the question, and the cause of the strike, from first to last, is for recognition of a union among the men. The question of hours was the direct cause, in other words, hastened the conflict; but that, as an out-sider, I didn't think that there would be any trouble about, for it was a matter about which there was very little at stake.[12]

The union in its growing sophistication had sent Boyce down to San Francisco to encourage a boycott of Wellington coal and to gain the support of Trade Unionists in the United States. From what the union president was telling the Select Committee about the causes of the strike it seems that the information he passed on to the the Americans was, at the very least, slightly inaccurate. A circular he distributed in San Francisco with the aim of persuading people not to buy Wellington coal certainly gave no indication that the matter was one of union recognition when it stated that, "These purse-proud aristocrats refuse to allow their miners to work for a living wage . . ." and described employers who were going to "carry into effect those brutal eviction scenes so prevalent in Ireland." The action of "enforcing" a boycott against Wellington coal, it was suggested, was necessary only because "Humanity demands that they (the Dunsmuirs) be taught a lesson of common decency." Another circular said "the firm demands that the men shall remain in the coal pits for TEN hours at a time, which is longer than human nature can stand without almost total exhaustion." The Council of the Federated Trades of the Pacific Coast declared a boycott on Wellington coal and, no doubt acting on information supplied by Boyce, advised that only New Wellington coal should be bought for there the miners received fair wages and were treated like human beings.[13]

The summer dragged on. There were no wages coming in except the monthly pittance provided by the union as strike pay. It was no more likely now than it had ever been that the Dunsmuirs would give way. Men who had been against the strike from the first and those who believed that it was all the fault of the union grew restless and on Saturday, August 2, a number of them entered the Wellington Mine and went to work. A rumour spread round the community that miners from outside the area would soon be coming to start work. At a hastily called meeting held the next day it was decided that there would be a boycott of the "scabs" and "blacklegs" and, in the hope that it would stiffen the resolve of those wavering on the edge of returning to work, a procession was planned to show that the main mass of miners was still very firmly behind the strike.

The procession was heavier on symbolism than violence but the potential for violence was clearly present. At the head of the procession a large Canadian flag was flying and behind it appeared a red flag said by the marchers to be indicative of danger which would, if necessary, be replaced by a black flag significant of destruction. The women participating wore black sashes. The *Colonist*'s correspondent noted, with apparent distaste, that most of the marchers were "foreigners," mainly Belgians, who were "tainted with socialistic ideas." The malign influence of the socialistic element he thought would have been of little importance had it not been for the liquor that was freely available. As it was, the worst of the strikers were trying to get the upper hand, and, if that happened, he considered it would be beyond the power of the inadequate police force to keep control. The morning procession contained only about seventy-five people; however another march, held in the afternoon, contained twice that number. An even larger procession was planned for the next day; the intention was that it would contain supporting elements from other union locals. Every effort was being made to shame the men who had returned to work.[14]

Police weakness would not be a problem: three riot-fearing magistrates had already requisitioned the militia. Lieutenant Colonel Holmes and sixty chosen men were on their way from Victoria under the cover of a secrecy imposed by a ban on telegraph messages. Each militiaman carried with him sixty rounds of ammunition as compared to the mere twenty rounds given to the troops attending the Wellington Strike in 1877.

The strikers were lining the road along which the "scabs" and "blacklegs" must pass at the end of their shift in the mine. Prominently displayed was the flag of the Knights of Labor and alongside was the flag of the miners' union. In addition a gallows was being carried from which hung a "blackleg" in effigy. The message was clear. So far the attempts to intimidate the men who had returned to work was having mixed results; while two men who had gone back to work had stopped working again, two others entered the mine after saying that they would have stayed out until the end of the strike if the attempt to frighten people had not been made. This procession was meant to be a major attempt to shame those who had returned to work, but things did not go quite according to plan, for the promised miners from the other locals failed to arrive. Perhaps it was just as well. Although there had still not been any open violence, threats were being used against the working miners. The *Colonist* saw the processions alone as "terrorism of the worst description arousing apprehension because many of the participants were pronounced socialists." It is unlikely that the men being called "scabs" and "blacklegs" were as frightened by the socialistic leanings of the mob as was the newspaper reporter.

But this was not to be a good day for intimidation. While the demonstrators were still waiting for the shift to end, news was passed to them that the militia had arrived by train from Victoria. In a body they headed down to where Lt. Col. Holmes was instructing his men on their duties and urging them not to offend anyone as this would avoid trouble. The troops were to wear only side-arms as they went about their duties and no one was to shoot except at his express command. It seems that Holmes was a man of good judgement. To the suggestion that he should act to disperse the crowd he replied that he saw no reason to do so as it was a peaceful gathering. The procession eventually broke up of its own accord after destroying the gallows and effigy at the request of the mayor.[15]

The procession of the next day was larger than ever, probably more to show that the strikers were not afraid of the militia rather than from any conviction that it would achieve any useful result. Once more the military was in Wellington to be used against striking coal miners if the need arose. The question to be answered was why was the militia there?

Immediately it became known that troops had arrived in Wellington, the elected representatives of the Nanaimo area sent a telegram to the Premier of British Columbia protesting the presence of the militia and urging its immediate withdrawal. In their opinion the sending of soldiers into an orderly and peaceful community was an unwarranted insult to the people which would in all likelihood provoke rather than prevent disorder. The representatives assured the Premier that they would hold the government strictly accountable for the actions of the militia. In a telegram remarkable for its brevity, the Premier icily informed them that their threat was superfluous as the government was always accountable. He made it quite clear that the provincial government had no control over the militia. As it happened not all the inhabitants of Wellington were complaining about the presence of the militia.

From seventy-three "workmen and and employees of Messrs. R. Dunsmuir and Sons, and residents of Wellington," the government received a petition, the original of which was "much soiled by having been carried through the mines having signatures placed on them." It was their desire to have protection provided by the provincial government if the militia was being withdrawn as was rumoured. All they wished was to be allowed to go and return from work peacefully without a recurrence of the scenes that had taken place on the first two days on which they had been exposed to the jeers and insults of the processions. The Attorney-General replied that he knew nothing of the removal of the troops as he had known nothing of their calling. The sending out of the militia was done at the instigation of Justices of the Peace who, it was assumed, would not permit them to be withdrawn until all danger was gone.[16] The federal government in Ottawa also disclaimed responsibility for supplying armed forces for use against civilians in a dispute with their employer. A protest from Mr. D. W. Gordon, M.P., drew from the Minister of Militia, Sir Adolphe Caron, a telegram which read: "Troops called out in aid of the civil power. Authority was from local magistrates, not from us."[17] The threat of the use of armed force was being used and neither level of government could apparently do anything about it. At whose instigation did the Justices of the Peace act? There is little doubt that the Dunsmuirs had been involved. It was said that James Dunsmuir, remembering the political fuss over the taxpayers having to foot the bill for militia use at Wellington in 1877, simply guaranteed payment of all costs. If that was so, then it was clearly implied that any man with sufficient wealth who could persuade three Justices of the Peace to sign a military requisition for militia assistance could, in fact, hire the militia and the government was powerless to prevent him from doing so. It seemed that Wellington and the military did not go well together, for each time the militia was called out the situation took on comic opera aspects. The troops were there and there they were likely to stay. The provincial government wanted them out and so did the federal government, but, as the troops had been sent for by the Justices to aid the civil power, only they could release them by their judgement of when that aid was no longer needed. No amount of persuasion could get the Justices of the Peace to declare that the emergency was over and eventually, on his own initiative, the commander of the force reduced its number to a sergeant and four men. Not until five months had passed were they withdrawn on the authority of the Minister of Militia in Ottawa.[18]

The strikers were fighting their battle on two fronts: not only were they standing out against the Dunsmuirs but they also had to contend with men who refused to join the strike. There can be little doubt that there were threats made to the men who had returned to work — probably more than ever became publicly known. There were also public insults and wordy threats, but there was no open violence serious enough to justify calling in troops. But then, it was not justification that the Dunsmuirs were looking for, only a renewed stream of coal from their idle shafts. Apart from the non-striking miners who wanted protection, there were also those in the community who had only praise for the way in which the mine owners had sent for the militia before violence had a chance to become established. However the purpose of the militia was not the same as it had been in 1877: then, in defence of the law, it had been interposed between the owners and the strikers, while in 1890 it was ostensibly to maintain peace between rival groups of miners. The industrial scene was changing rapidly towards the end of the nineteenth century, unions were becoming militant and were proving to be no more tolerant of workers who would not follow the union line than they were of employers who would not meet their demands. They complained, and often with good reason, of the inhumanity of employers to workers but were not models of humanity themselves. There is no telling how much worse the Wellington Strike of 1890 might have been if the miners had taken literally the words of a union journal of the time which read:

> . . . a scab is a specimen of humanity calling himself a man, but is a traitor to his fellow-men. He is a thing to be despised, to be hated, and to receive the contempt, the ridicule, and the most intense hatred of all honourable men. . . . he is a nondescript; neither man nor beast, but is a kind of reptile that walks on two legs but would appear to much better advantage crawling on his belly . . . From a scab may the good Lord deliver us.[19]

So sure were they that the means justified the ends that the name of the Lord was invoked in advocating inhumanity to other men to a degree that would have rightly appalled them had it been practised towards them by the employers. Early in 1891, when the strike had been going on for several months, a group of union supporters, who worked at other mines in the district and were not themselves on strike, went some way in applying the quotation in a way that brought howls of outrage from letter writers and also caused the government to appoint a Select Committee of Investigation.

"Truth" in a letter to the *Colonist*[20] professed himself disgusted by the actions of union supporters. They had committed "one of the most ruffianly and disgraceful scenes ever witnessed," which caused him to wonder if he was living in a civilized country. If this was unionism, then "the less of it we have in British Columbia the better." Why could the strikers not accept that they had been ignominiously beaten by the owners and acknowledge that they had fought a battle and lost? Two days later a letter signed "Spectator"[21] supported the views of "Truth" and described the unionists as "people whose judgment has been warped by agitators, who have overwhelmed them with their irrational balderdash and confident, unblushing falsehood." Even savages, "Spectator" pointed out, will allow their enemies to bury their dead in peace — in contrast the union hounded to the grave those who did not share their views.

Ellis Roberts had been a miner. Like many others before him he had died of injuries received in the mine — what set him apart from the others was that he had died in the mine during the strike called by the union. In many union men his death evoked neither sympathy nor regret; had he been on strike as he should have been instead of working below as a strike-breaker he would still have been alive. Served him right for being a "scab." The body of Roberts was to be taken from Wellington to Nanaimo for burial. The cortege of sleighs and wheeled carriages, containing fifty or sixty non-strikers, was drawn up ready to leave. At almost the same point the usual strikers' procession was being formed with flags flying. It seemed that there might be trouble, but there was not. In fact it was later claimed by the strikers that the flags were dipped in respect to the dead man as the hearse passed. There was no sign of trouble on the way to the cemetery either; however, while the service at the graveside was going on, one man heard "unearthly sounds" coming from the bush while another heard a loud voice say "Put the son-of-a-bitch in the hole." The minister engrossed in reading the funeral service heard "no unseemly interruption."

Seldom were so many "scabs" assembled together in one place and for some the opportunity for mischief was too good to miss. Going back through Northfield the procession was pelted with snowballs by women and children urged on by male bystanders; an effigy of a "blackleg" with a black crepe band around an arm was held up to them; "Truth" reported "a torrent of blackguardism, among the choice epithets being 'scabs' and 'blacklegs'"; and two women displayed their black-stockinged legs. "They showed their legs to us and they went rather too far, I should think — above the knee," as one witness described it. One of those attacked said, "They was not so very vicious," and the Select Committee appointed to investigate the matter agreed, finding "the conduct of the people of Northfield to the non-union men returning in procession from the funeral was unseemly and insulting."[22] Insulting was the important word, for as "Spectator" had warned in his letter "people cannot stand being insulted while going about their ordinary business. Even a worm will turn if trampled upon." His simple message was that, if the insulting behaviour of the union did not stop, there would be violence done to them. The danger was very real.

When news of the funeral outrage got to the miners who were working underground at Wellington it was only through the intervention of men with authority among them that the more indignant were prevented from taking pick handles to those who had caused them to be insulted.[23] There is little doubt that they could have inflicted considerable damage for by that time there was a rump of about 150 union hold-outs on strike, while in the Wellington mines there were at work almost 500 non-union miners. The revulsion against the attack on the funeral party and the threat of still more violence brought official action. Within days men found guilty of intimidation in the preceding October but not sentenced were sent for by the court for sentencing. And, in an effort to stop the strikers' processions which had gone on throughout the winter on Mondays, Wednesdays and Fridays, other men were arrested and sent for trial charged with openly defying the ban on processions imposed by the Chief Justice. It was ironic that the action which had finally led to the strikers being discredited was one carried out by miners from outside the area; just as it was outsiders who had helped to bring about the strike by their votes at the union meeting of almost a year before.

By mid-April the *Colonist*, never a paper to back strikers over the mine owners, was announcing with glee that the strike was over and rubbing in the message by enlarging on the booming state of Wellington. Five hundred underground workers were sending up 1,200 tons of coal each day. The new men brought in were delighted with the money they were able to make in the mines and were determined to stay in the town. As all the Dunsmuir houses were occupied a new town site was being prepared; on it would be built, so the newspaper said with figures more sweeping perhaps than accurate, two or three hundred additional houses. The editor could not forego the chance to comment that the strikers were now wondering why they had ever gone on strike in the first place.[24]

So another strike ended with the Dunsmuirs as clear winners. But why was it that they were continually embroiled in strikes while the mines of the neighbouring Vancouver Coal Company seemed immune from such troubles? A good part of the answer may lie in the the personalities and management styles of the two companies. The most recent strike was about union recognition, and so had to be called against the Dunsmuirs because the other company recognized the union. The Dunsmuir brothers were "chips off the old block"; they were of the mine owning family; they arranged their affairs to suit themselves and to bring in the absolute maximum in profit for their own benefit; they would tolerate no interference. The Vancouver Coal Company was an absentee owner with its main office in London; its affairs in Nanaimo were controlled by a resident manager. When the Dunsmuirs saw something in the mine that displeased them, they could, and did, as the ultimate authority, change it without pause for reflection. When Samuel Robins, as manager of the other company's mines, wished to change things, he was not the ultimate authority and had to pause and reflect on the likely reception of the change by his employers in London. His was not the luxury of autocratic control; he lacked the almost "god-like" attitude towards employees which seemed almost to have been the birth-right of the Dunsmuirs.

The union was recognized in the mines of the Vancouver Coal Company because Samuel Robins approved of unions. He understood that men could suffer from the arbitrary whims of those over them which left miners to either swallow their pride and submit or be kicked out of the mine to face starvation. In his opinion the union influence was good, for it gave a man the feeling of self-respect; he knew that there were ways of appealing arbitrary rulings and this lessened his fear of injustice. It was his opinion that strikes were inevitably the result of the "failure of temper or absence of reasonableness either on the one side or the other" and frank discussion with the union could often lead to a fair and just settlement of differences. It also helped that, unlike the Dunsmuirs, Robins was opposed to using Chinese workers underground and used them on the surface only because his company would otherwise be unable to meet the competition of coal from Washington Territory which entered the San Francisco market free of duty.[25] It was Robins who had immediately removed the Chinese from underground after the Nanaimo explosion, while the Dunsmuirs temporized and then, after removing them from the Wellington mine, continued to use them underground in the developing coal-fields of Comox. There is little wonder that Nanaimo celebrated an annual "Sam Robins Day."

And yet the Dunsmuirs still managed to get all the miners they needed to work their coalmines and the line between the companies must not be too firmly drawn. They both ran coal mines in the same age and conditions were probably not that far apart. Robins encouraged home and land-ownership while the Dunsmuirs provided low cost rental housing. Whatever might now be thought of the advantages and disadvantages of either, the working class of the late nineteenth century did not expect to own a home. As a large proportion of the population would at any time have been fairly recent immigrants with little money available, it might well have been easier for them to go into the Dunsmuir mines knowing that they would then be provided with a house at a very modest rental.

The strike had been a dismal failure for the miners. They had sent out their call to unite in the cause of union recognition and had proved only that they were not strong enough. They could not, as yet, persuade enough men to rally round the flag of unionism; until that happened they would never be able to bring enough pressure to bear on the Dunsmuirs to make them pay attention to the union. They had taken only a first faltering step along a road that was to be long and bitter.

9. The Century Ends

With mechanical precision the hands of the clock moved to midnight. A year and a century had died and another year and another century had been born. The story of coal mining on Vancouver Island was half told and already its beginnings were becoming lost. Fifty years previously the island, Britain's most recent and least important colony, had been almost empty of white inhabitants. The Indians occupied the land as they had always done and to trade with them the Hudson's Bay Company kept a handful of men in two forts. Fort Victoria, the main trading centre and capital of the colony, was at the south-east corner of the Island; an island length away to the north, Fort Rupert stood guard over what the Company hoped was a valuable coalfield. At the century's turn the city of Victoria masked all trace of the fort about which it had first been built and at Fort Rupert only a charred ruin still stood as a reminder of what had been.

Around Nanaimo, the site of the second and successful find, coal had brought growth and permanence. The town had prospered and settlements had grown up round about as new shafts were sunk into the coal seams to the north and south. Separate and distinct, another coal field was being developed to the north in the Comox Valley and there too before the century's end new towns were being created. With well over three thousand men toiling in and about the coalmines of Vancouver Island the future seemed assured. However, as early as the end of the first fifty years, there were foreshadowings of the doom which must inevitably come to areas dependent on a single, finite, natural resource. The town of Wellington came into being after

Robert Dunsmuir's coal find at Diver Lake in 1869, and it died after only thirty years through lack of coal. Its demise was as rapid as its birth. The *Nanaimo Free Press* noted in 1876 that where only six years before there had been wilderness a village now flourished; so advanced were its mines that the daily production of two hundred tons of coal could be moved to the dock by an engine capable of moving twenty tons at a time at the remarkable speed of eight miles per hour.[1] At the end of the 1890 Strike everything was looking so rosy that two or three hundred houses were required to house the extra miners needed to increase production in mines in which the coal reserves were said to be almost inexhaustible.[2] In its short and turbulent history Wellington grew until it had five thousand inhabitants; in a burst of civic pride the town was incorporated in 1894. Five years later James Dunsmuir announced that because coal supplies were becoming exhausted the Wellington Collieries would be closed within two years. It was in the first year of the new century that the last ton of coal was mined from the infamous Number 5 Shaft, site of the spectacular disaster of 1888.[3] If the coal was finished, the town was finished; it had no longer any reason to exist.

One man, Robert Dunsmuir, had called Wellington into being and his sons, by their own decision, destroyed it. For the most part the buildings of Wellington were dismantled and moved a few miles south of Nanaimo to Dunsmuir's Extension Mines. No sooner were the people settled into Extension than James Dunsmuir passed the word that they were again to move. This time the move was to Oyster Bay a dozen miles away — or what had been Oyster Bay before Dunsmuir decided that its name should be changed to Ladysmith in celebration of the lifting of the Siege of Ladysmith by Empire troops fighting in South Africa. Many of the miners simply stayed where they were in Extension but took the precaution of walking down the railway line towards Ladysmith and boarding the train so that they could be seen arriving in Extension by train to work their shift in the mine.[4] There is no better illustration of the power of the coal bosses than in that they could create and destroy towns, move and rename them at a personal whim and not only tell employees how they should work but where they must live. Perhaps the resistance of the miners to such dictatorial behaviour was an indication that they would not put up with it much longer. It was, at the very least, a sign of changing times.

Coal mining on Vancouver Island had its beginnings in the Victorian Age; its economics and attitudes were a reflection of that rough, tough period when men spoke directly and said what they meant without evasion. Complex as the technology was at the turn of the century, in contrast to earlier times, there remained an essential and basic simplicity in the relations between men. The boss was the boss, and he knew it and invariably acted in the light of that knowledge. Workmen knew who was boss also and, even though they might not like him very much, they usually accorded him great respect — a respect based partly on fear and partly on their envy of his open display of wealth and success. The distinction of class was obvious and clear; even if the workmen resented the superiority of the boss they were not above creating distinctions of their own — a vast difference could be expected in the respect given by them to a foreman, a skilled worker and a simple labourer. But even the lowest among them had, in their own eyes at least, the saving grace of being white and, therefore, having lesser beings to look down upon. The Indians pointed out the coal sources and having given up their coal-bearing lands to the avarice of the coal entrepreneurs, were of no further value and not worthy of consideration. The Chinese were of greater interest, for they were capable of doing work for the bosses at low rates of pay and were, as a result, also capable of giving economic competition to the white workers — a competition not making them in any way equal to the whites as far as anyone could see. The modern terms of class distinction and racial discrimination were not bandied about; the reality of differences was to be seen every day, arranged as God in his wisdom no doubt wished it to be. The arrangement, at least as far as the racial aspect applied, had the blessing of the white coal workers who undoubtedly approved of what they saw as their God-given superiority.

That Chinese should be involved at all shows that the story of coal on Vancouver Island cannot be explained satisfactorily by Victorian economics and attitudes alone. The unique flavour of the tale owes as much to the pioneer conditions of the island, its North American setting and the presence of the Chinese, as it does to the ideas of the age brought in as part of the unseen baggage of so many of the immigrants who worked coal and those who became rich through its exploitation. The story of the first fifty years of the Vancouver Island coal industry is, in the main, one of abrasive contacts between the mine bosses and the white coal workers with the unfortunate Chinese caught between.

The pattern of relationships between employers and workers was established early and was not unusual for the age. The Hudson's Bay Company wished to profit from its coal discoveries and was not prepared to put up with nonsense from miners in the achievement of that aim. For their part the miners were not prepared to put up with the autocratic behaviour of the company's officers and were prone to assert their point of view through strikes. At Fort Rupert this had led to fines and to imprisonment in chains on a sparse diet of bread and water before the miners had deserted the insubstantial coal workings. Later, at Nanaimo, strikes had led to a reduction of the wages of miners who had dared to try their luck in mines in the United States and had then been forced to return to Nanaimo when faced with the threatened eviction of their families from company-owned houses. In a very few years the pattern of action and re-action had been firmly set — there would be friction leading to strikes, to violence and to evictions while a scrap of coal remained to be mined on the Island.

When the Vancouver Mining and Land Company bought out the coal interests of the Hudson's Bay Company it also had weapons to be added to the armoury for use against awkward and stubborn miners. This company was the first to use "Cards" in the newspapers outlining for the general public the rightness of its views as opposed to those of the miners and it was also the first to use Chinese workers, not only to cut costs, but also as a strike breaking tool. It was left to Robert Dunsmuir to add the final touches of oppression by using evictions on a large scale, by hauling into court on a variety of charges strikers who had particularly displeased him, and, worst of all, calling in the troops to fight his business battles for him on trumped up concerns for public safety. To his sons, James and Alexander, went the task of refining and making more efficient the weapons to be used in safe-guarding the interests of mine owners.

The employers considered themselves to be public benefactors for without their efforts there would be no work. The business risks they had taken and the skills they had applied were the foundation of the growth of Vancouver Island; before the exploitation of the coal deposits there had been little enough reason for it to exist let alone prosper. For their efforts they expected an adequate reward. The business outlook of the age led them to use whatever means were available to make the greatest possible amount of money in the shortest possible time. They clearly believed that they were superior to common workmen, who were of value mainly as tools to be used and discarded at will, and from all the outward signs it was difficult to argue that their assumed superiority was false. Often they had to take action against the men in their employ but not, they would have argued, in a vindictive spirit: rather they saw it in the manner of a father obliged to punish wayward and not very bright children for their own good — punishment meted out in sorrow rather than anger, paternalism personified.

The employers were not without their own real problems, apart from what they saw as the intransigence of workers. Over a fifty year period the mines had grown beyond recognition in men employed, coal produced and profits earned, but the growth had been far from steady. The major market of San Francisco was very variable in its demand for coal and the coal producers of Vancouver Island could in no way influence the fluctuations. They also had to pay a tariff on each ton of coal they sent into the United States which made it very difficult to compete with domestically-produced coal. In the first eleven months of one year Robert Dunsmuir, only one of the producers of Vancouver Island coal, was reported to have paid $210,000 in tariffs.[5] By the end of the century there was another problem which threatened to have serious consequences. That was the introduction of oil as a fuel; the switch from sail to coal by the ships of the world had made the coal industry and a further change from coal to oil could break it. To compete effectively the coal industry of Vancouver Island had to produce more cheaply, either by increased mining efficiency or by limiting production costs. The broken nature of the coalfield made difficult the use of mechanical mining methods which had the potential to bring increased efficiency and so the pressure was passed to the coal workers. Thcy were expected to produce more coal while the total wage cost was kept down by the extensive use of Chinese labour.

Opposition to the dictatorial ways of the mine owners and managers was certainly well established by the time the new century dawned and it was being formalized by unions of increasing power and vociferousness, but which were not yet strong enough to cause the employers problems beyond the hope of resolution. There was no lack of clouds on the far horizon but Joan Dunsmuir still occupied her castle; James Dunsmuir had built for his family a house far surpassing in size and luxury Craigdarroch Castle; and, not far from San Francisco, Alexander Dunsmuir had built for Josephine Wallace a mansion in extensive gardens set in a three hundred acre estate.

The latter half of the nineteenth century was a teeming compound of dissatisfaction and hope; travel was slow, difficult, and uncomfortable, but millions of people were on the move searching for a better life. The coal discoveries on Vancouver Island made it certain that some part of the immigrant flood would flow to the northwest coast to fill the empty land. The earliest of the pioneers spent at least half a year on the voyage from the British Isles; a voyage as long and dangerous as could be found anywhere on the world's oceans. As time passed others sailed to Panama and having crossed the isthmus re-embarked on the Pacific Ocean side to sail north to Vancouver Island. Later immigrants were able to travel to the the Atlantic shore of the United States and then head west across the continent on the newly built railroads, sometimes having to traverse uncompleted sections by stage coach. Finally, when the century was not much more than a decade from its end, newcomers were able to disembark from steamships in Canadian eastern ports and take a direct rail journey to the west coast by the Canadian Pacific Railway. The discomfort of travel lessened over the fifty years, but the journey was never easy. Diverse as the immigrants to the Vancouver Island coalfields were, they had in common dissatisfaction and courage: dissatisfaction with life as they had known it and the courage to take a large step into the unknown. They were a turbulent mixture, rough in manner and largely uneducated, yet with a knowledge and awareness of the world far beyond what could be expected in a place so geographically remote. They had all known other lands and arrived with attitudes and perceptions honed in more advanced societies. Through time the edges of old attitudes and beliefs would be modified by their new environment, but always successive waves of immigrants would bring news and new ideas to keep that awareness of the outer world alive.

Coal miners were different; tough, independent and with a prickly individuality that made them difficult to get along with. Their working conditions as they hewed coal in the confined spaces of the dark, damp and filthy mines were beyond the imagination of most other workers. Few men who spent any length of time below ground in the mines escaped without scars to show for it; always in the background lurked the fear of disaster from roof falls, from the insidious and silent gas or from a sudden, cataclysmic explosion. It was a hard life but the miners who worked like moles in their underground tunnels walked on the surface like men: individuals with strong personal opinions who would not easily yield. Their strength of opinion and character could, at times, also be a tragic source of weakness. All the miners worked in similar conditions and so had reason to complain of the same things and yet there was often a marked diversity of opinion. The seams of the island coal mines were in many instances fractured in such a way as to make the use of machinery impractical for mining and more often than not it came down to one man in his own workspace using his own muscles and brains to work in his own way. The common thread of experience promised to bring united action; however working against that tendency there was often a mulish individual streak which, although promising stiff resistance to the bosses, brought with it an almost certain guarantee that there would be no easy agreement with the ideas of other miners. They had a union tradition and they had unions, but often what should have been sustained resistance to the employers for the good of all miners collapsed on their own inability to agree among themselves. A tragic flaw and one the employers well knew how to exploit.

And then there were the voiceless men who had also travelled far. It was a long way from the coal mines of Vancouver Island to where the Pearl River, having drained much of southern China, flowed through a broad gulf into the South China Sea. Over the years a delta had grown where the silt-laden waters were slowed by the bulk of the sea. On the delta lands of the fourteen counties of Kwantung Province too many people struggled to exist on too little land. Rents were high; money to buy fertilizer scarce; government was chaotic, oppressive and corrupt; there was banditry and there was warfare. Conditions were bad and yet stark poverty made it almost impossible to leave and steel-strong family ties made it unthinkable that those who might leave to work overseas would not eventually return. At the head of the delta stood Canton,the capital of the province, which had always been one of the great seaports of China. In spite of the difficulties, from the crowded lands accessible to the sea and to seaports a tradition of migration to other parts of Southeast Asia had grown. Forced contact with Europeans in the mid-nineteenth century extended the migration pattern to lands settled or controlled by those of European descent. The Chinese were soon to be found in Africa and Australia; in the gold fields of California and those of the Fraser River, and in the coal mines of Vancouver Island.

Vancouver Island was a place containing very little that the Chinese recognized or understood. They were there and they were not there: useful as manual labourers but treated by the white population as though they had no real existence. There was no possible way that the Chinese could become a part of the society they served and so they were thrown back on their own deep sense of kinship. The basic social organization they had always know was that of a group of families with a common ancestor living together in a village for mutual support; this deep-rooted pattern was carried to Vancouver Island. In the early 1880s, as an example, 149 of the 245 Chinese resident in Nanaimo were from the T'ai Shan province of the Kwantung Peninsula and of the total almost one half shared the single surname "Ma."[6]

The Chinese had known poverty at home and they had not entirely escaped it in Vancouver Island. It is possible that they were financially better off than they had been at home; nevertheless the fact remained that they were being paid only one half of what a white worker could expect to be paid for doing the same work. Simple and naked exploitation was their lot. They had known limitations on their freedom at home and they were still not free. The fare to North America had been beyond their means and they had been financed by "coolie brokers" who had to be repaid and with interest. They were trapped by their debts and prevented by low wages from easily freeing themselves while at the same time trying to send money home to China for the support of families left behind. They had every reason to be docile workers, unlikely to strike on their own account and useful as strike-breakers when the white miners walked out.

One of the main complaints the white population had against the Chinese was the way they lived in over-crowded and unsanitary conditions. With low wages, debts and the obligation to send money home, they had little choice. The whites who complained about the Chinese did nothing to help them improve their living conditions. The Chinese were disliked simply because they were Chinese, a pervasive dislike "firmly rooted in the rich, hard clay of public opinion."[7] In the mining districts this dislike was intensified by the massing of Chinese workers in a way that was to the obvious economic disadvantage of white workers.

The employers' attitudes to the Chinese were simple and understandable: the Chinese were cheap to employ and their use enabled greater profits to be made. There were no moral considerations, no humanity; it was a simple business affair and as such required no justification. Whether or not the Chinese were desirable as people did not enter into the picture. The attitude of the workers was quite the reverse of that of the employers, but was also simple and not to be confused with humanitarian considerations. The Chinese took jobs that should have gone to white people and also held down wages — the Chinese must go. The mine workers and their unions missed no opportunity to support any form of proposed legal discrimination whether it be by taxation, by denial of voting rights, by direct expulsion of the Chinese already on the island or by banning of Chinese immigration. But in spite of the fears of violence expressed to the various commissions, there is little evidence to suggest that the Chinese of the coal mining areas suffered much from physical attacks up to the end of the century. The Nanaimo newspaper carried occasional reports of interference with the Chinese which indicated the kind of teasing behaviour associated with ill-mannered adolescents rather than determined attacks by men out to do real damage. These isolated and infrequent attacks were invariably condemned by editorial comment. The real cruelty to the Chinese was far worse than physical violence in that it denied their humanity. They were defined in terms of use to the coal owners and, wordy though their opposition to Chinese labour might have been, the miners shared in the guilt for they also were not above employing Chinese assistants and paying them the same starvation wages as did the owners.

The mine workers of Vancouver Island seemed incapable of recognizing that the Chinese, caught between the white workers and the white employers, were, like them, part of the world-wide battle between capital and labour. The whites had a choice: accept or reject the Chinese. They chose rejection and as a result became ineffective in the fight against the mine owners because alone they had insufficient numbers and power. Had they chosen to accept the Chinese into the unions and fought for pay equality they might have served their own ends better. But there was no hope of that.

In 1888, as the last of the bodies were still being recovered after the Wellington Mine Disaster, an article appeared in the *Colonist* commenting on the decision of a Knights of Labor branch in the United States to accept Chinese workers as members. The writer saw this development as not only strange but startling and regretted that in accepting the Chinese as equals the Knights had not looked sufficiently far ahead. He saw no merit in the intention of raising up the Chinese to pay equality with the whites as a means of keeping them from being strike-breakers for giving them more money would only result in greater floods of Chinese immigrants and as they could always be expected to live on less than the white worker there would be pressure for wages to go down again and the Chinese would become the preferred workers once more. There was only one way to deal with the Chinese, the writer contended, and that was by complete prohibition[8]. There were many factors at work obscuring the situation and the white workers were much too close to the problem to see it with the clarity always claimed by those who come after.

And what did the Chinese have to say about all this? Who knows if they were even aware of most of it? Two commissions of inquiry were set up to investigate the Chinese situation in British Columbia at the century's end and no single Chinese immigrant was heard. They had no voice. The problem was no nearer to a permanent solution as the year 1900 arrived. However there was a complication arising and that was to be found in the increasing number of Japanese immigrants entering the mines, immigrants with a home government not content to be as silent about open racial discrimination as the Chinese government had always been.

Very few of the pioneers of the Vancouver Island coal industry lived to the end of the nineteenth century. Dr. William Fraser Tolmie, the young physician and fur trader who from his post at Fort McLoughlin first reported the Vancouver Island coal deposits, was gone after a long and useful life. Dr. John McLoughlin, the chief trader of Fort Vancouver, to whom Tolmie made his report and who arranged for the steamship *Beaver* to call in and examine coal seams, stayed in the "Oregon Territory" after the boundary was drawn and after his death was honoured and respected as the "Father of Oregon." The first governor of the island colony, Richard Blanshard, who visited the troubled Fort Rupert in the days of the Muir's explorations of the coal deposit, stayed in his post for two miserable years before leaving for more congenial places. James Douglas, founder of Fort Victoria, second governor of the Colony of Vancouver Island and later also governor of the newly created mainland colony of British Columbia, died and was buried in Victoria. His enthusiasm for developing the coal finds about "Nanymo Bay" had been instrumental in giving the Vancouver Island colony an economic reason for existence.

Of the Fort Rupert miners, John Muir, the leader of the original group and also one of the first explorers of the Nanaimo coalfield, died in 1883. It was from the early workings at Nanaimo that he had retired to Sooke to prosper as a farmer and lumberman, and it was from Sooke that he carried on a life of public service, being first appointed Justice of the Peace and then going on to to serve for ten years in the Legislature. His main claim to historical fame in British Columbia, apart from his work as a pioneer miner, is that he was a member of the first legislative assembly of Vancouver Island. Others of his party led less distinguished lives. John Muir's nephew, Archibald Muir survived him by only four years dying in the Nanaimo mine disaster of 1887. He in turn, had outlived by many years the truculent John McGregor who along with Andrew Muir had defied Captain McNeill and his minions at Fort Rupert and ended up imprisoned and in chains on a diet of bread and water.

The Brierley miners who arrived in Nanaimo as the second wave of settlers suffered a variety of fates. George Sage and his wife were drowned in Sage Bay off Newcastle Island in Nanaimo harbour due, it was claimed by the over-imaginative, to being dragged into a cave by an octopus.[9] Edwin Gough was involved in a accident in a Washington coal mine and taken to the nearest doctor by canoe. The doctor, who was no nearer than Victoria, promptly amputated the injured leg. The experience is said to have changed Gough from a gay, spirited person to one with a sad air of despondence; understandably so after his harrowing journey and its surgical aftermath carried out in the rough and ready fashion of the day. He later entered the hotel business and died in 1875.[10] His companion of the outward voyage from England on the *Princess Royal*, John Meakin, died in the same disaster as did Archibald Muir.

There were a few aging survivors left at the end of the century who could recall those first mining days. Dr. John Sebastian Helmcken, the Hudson's Bay Company doctor set by Governor Richard Blanshard the impossible task of refereeing the dispute between the miners and the Company officers at Fort Rupert, still lived in Victoria in his seventy-fifth year and still had twenty years to live. While following a medical career he had also served as a politician, being the Speaker of the first Legislative Assembly and one of the three representatives sent to Ottawa to negotiate the entry of British Columbia into the Canadian Confederation. With only a few months left to live, Joseph McKay was also a resident of Victoria. He had arrived on Vancouver Island as a youth of fifteen years of age and by the time he was twenty-three was the overseer of the Company's attempt to mine the Nanaimo coalfield, spurred on by the steady stream of letters flowing from the pen of James Douglas. He also became a member of the first assembly. John Thompson, reputed to have been the first of the Brierley Hill miners to set foot on shore at Nanaimo in 1854, would live until 1910. His claim to fame seems to have been in the size of his family. On the death of his first wife, whom he had married immediately before leaving the "Black Country" for Nanaimo, he returned to England to choose a new bride and ended up the father of twelve sons and three daughters.[11]

Of those pioneers gone and those remaining, two were of outstanding importance: Robert Dunsmuir died a decade before the end of the century, but his wife, Joan, lived on. Robert Dunsmuir's story was the stuff of which Victorian dreams were made; a man using his talents and hard work to attain a deserved reward of respect, wealth and power. It was almost the classical "rags to riches" story, but with a touch of mythology that would increase with the passing of the years, for it was not quite from rags that he started his rise to wealth. He had undoubted ability, aided by the luck to always be in the right place at the right time needed by even the most able, and above all he had that streak of ruthless determination to achieve his goals. If there was blood on the coal from which he took his profits then that was only to be expected when men worked underground — there had always been blood on the coal. If he paid the Chinese in his employ wages that caused them to live in poverty and treated his white workers with a stern paternalism when they transgressed, those were the simple

Fairview, the Victoria house where Robert Dunsmuir died in 1889.

realities of the social and economic attitudes of his time. There is no doubt that Robert Dunsmuir was believed by many of his compatriots to be the man to whom most was owed for the economic progress of Vancouver Island, but he was also believed by many to have been responsible for creating much of the friction that existed between employers and workers in the coalfields. It is at least arguable that the prevailing attitudes of workers and employers in British Columbia can be traced to Dunsmuir foundations. On the death of Robert Dunsmuir, control of the coal empire passed to his wife. As the century ended Joan Dunsmuir sat in her baronial mansion in the midst of her Victoria estate, the only remaining pioneer to have power and influence left to wield. She, the Scottish village girl who had become within her confined world an undisputed queen, was reputed to be harder and more determined than her husband had ever been.

From the time of their father's death in 1889, James and Alexander Dunsmuir were anxious to buy the company from their mother. Their lack of success was tied up with "that woman" as Joan Dunsmuir referred to Alexander's consort in San Francisco. Quite simply, if Alexander owned a half share in the company and he then died after marrying Josephine Wallace, she, as his wife, would very likely inherit half the company. That the woman with whom her son had "lived in sin" should be so favoured while her own daughters were cut off from the source of the Dunsmuir wealth was unthinkable and for more than ten years after her husband's death Joan Dunsmuir refused to give in to the pleas and arguments of her sons.

Alexander Dunsmuir was trapped between the pressures of wanting to marry Josephine Wallace and the fear that, if he did in the face of the fierce opposition of his mother, he might be in danger of losing his livelihood. From that ever present dilemma and possibly because of flaws in his character and disposition, he began to drink to excess. He became changeable and moody, almost incapable of making business decisions, and, when he had made them, unable to hold to them. He was increasingly reliant on his brother for establishing company policy. It was mainly James who continued to press Joan Dunsmuir to sell the company to them in meetings that grew ever more stormy. She finally agreed, but only on the condition that Alexander would first make a will naming James as his beneficiary, so effectively cutting his wife off from directly benefiting from ownership of the company. Alexander complied and the deal was made. That done, he and Josephine Wallace married on 21 December, 1899. Though they had been together for many years, the marriage lasted only forty days, for at the end of that time Alexander Dunsmuir died of alcoholic dementia. His wife got the house he had built for her and James became the sole owner of the Dunsmuir coal empire.[12]

As the new century began the company was once more under the control of one man, but that did not mean a return to the days when Robert Dunsmuir ruled alone. James Dunsmuir was like his father in many ways and not least in his business methods, but times had changed and so had the Dunsmuirs. The family, that had arrived in the steerage section of a wooden sailing ship and survived its wreck at the end of a six month journey, and then gone on to live in pioneer shacks in the same conditions as everyone else, now lived in luxury with aristocratic connections. It was not so aware, or perhaps was less willing to admit, that it had once been poor. The struggle for success was long over and wealth had imposed on James Dunsmuir

Craigdarroch. Joan Dunsmuir moved into the castle one year after her husband's death.

obligations which he felt he must meet. In 1900 he became, with reluctance and no great talent for the job, the Premier of British Columbia. In the next year he went to London to attend the coronation of Edward VII, delaying his return until early 1902 in which year he also resigned as Premier. It seemed that his taste for responsibilty was becoming diminished, although his business sense remained keen, for in that same year he bought out the shares of his San Francisco associates in the Comox Mines and the Esquimalt and Nanaimo Railway and then three years later sold the railroad alone to the Canadian Pacific Railway for three million dollars. He was starting to disengage himself from his business interests.

It would be foolish to claim that the coal industry changed at the turn of the century with the precision of measured time; however, clustered around that point there were changes of great importance. One important change was towards greater size and impersonality. Two major companies would dominate the future direction of the coal industry, but they were not the two companies dominant in the pioneering days; the union would still exist, but it also would be larger and less personal to the men who were its members.

As the immediate successor of the Hudson's Bay Company, the Vancouver Coal Mining and Land Company had been the foundation of the coal industry. In the tradition of British imperial business, the company was controlled by absentee owners; yet manager Sam Robins gave it a more co-operative character. In the early years of the century the company was bought out by the Western Fuel Company, another absentee owner based in San Francisco, and Samuel Robins left to spend his retirement in England. With Robins gone the old attitude of co-operation with the miners was replaced with an avaricious and anti-union outlook which could easily have matched that of the Dunsmuirs. Within a short time the new company was embroiled in disputes with its workers who had fighting for them the militant and radical Western Federation of Miners — an "absentee" union based in the United States. In 1910 the Dunsmuir connection with coal mines ended after a spectacular forty years, when James Dunsmuir sold out to a company that was allowed to retain the family name as a part of its official title: Canadian Collieries (Dunsmuir) Ltd. Retention of the name was unfortunate for it carried with it reminders of all the past injustices the mine workers believed they had suffered at the hands of the Dunsmuirs. The future actions of the new company would do little to wipe out the workers' memories of injustice and much to keep them fresh.

Dunsmuir's sale of his coal interests had completed the transition to more modern forms of industrial organization. The game now had new organizations and new players ranged against each other; yet, dead though the pioneer age might be, the attitudes fostered in that age were still alive. Coal had given to a few riches and power and influence greater than they could ever have imagined possible, but the mass of men who lived from coal still toiled for an uncertain weekly wage in conditions that could never be pleasant and were often dangerous. There had been strikes and there had been violence from the very start of mining. Over the years this had grown into a sullen and bitter distrust between owners and workers. Attitudes had been formed, hardened by reality and added to by myth and legend. The Chinese had been shamelessly exploited by the employers and flagrantly discriminated against by the white workers. All the problems had been created in the pioneer period; none of them had been solved. They remained like jagged, open wounds ready to fester at the slightest provocation. There had been hard and bitter times, and there would be more. Things could be worse, and they would be. Much worse.

Source Notes

Chapter I

1. William F. Tolmie, *The Journals of William Fraser Tolmie: Physician and Fur Trader* (Vancouver: Mitchell Press, 1963), p. 29.
2. H.H. Bancroft, *History of British Columbia* (San Francisco: 1887), XXXII, p. 158. Quoting John Dunn, *History of Oregon Territory* (London: 1844).
3. Stewart H. Holbrook, *The Columbia* (Toronto: Rinehart and Co. Inc., 1956), p. 66.
4. Anon., "The Founding of Victoria," *Beaver*, (March 1943), Quoting Simpson to the Governor and Committee of the Hudson's Bay Company, Honolulu, 1 March 1842.
5. *Ibid*. Quoting Minutes of the Northern Department of Rupert's Land, Norway House, 28 June 1842.
6. A.S. Morton, *The Canadian West to 1870-71*, 2nd. ed. (Toronto: University of Toronto Press, 1973), p. 750.
7. *Ibid*. p. 753.
8. Barry M. Gough, *The Royal Navy and the North West Coast of America: 1810-1914* (Vancouver: University of Vancouver Press 1971), p. 100. Quoting Gordon to Duntze, 14 January 1846.
9. *Ibid*.
10. *Ibid*. p. 102.
11. *Ibid*. p. 101.
12. J.H. Kemble, "Coal From the North West Coast 1848-1850," *British Columbia Historical Quarterly*, April, 1938. p. 12.
13. *Ibid*. p. 126. Quoting Simpson to Board of Management, Fort Vancouver, 13 October 1848.

Chapter II

1. Blanshard to Grey, 26 December 1849.
2. F.W. Howay, *British Columbia: The Making of a Province* (Toronto: Ryerson Press, 1928), p. 105.
3. W.E. Ireland, "The Appointment of Governor Blanshard," *British Columbia Historical Quarterly*, July, 1944, p. 213.
4. Blanshard to Grey, 8 April 1850.

5. E.C. Guillet, *The Great Migration*, 2nd. ed. (Toronto: Toronto University Press, 1963), p. vii.
6. *Ibid.* p. 18.
7. M. MacFie, *Vancouver Island and British Columbia* (London: Longman, Green, Longman, Roberts and Green, 1865), p. 145.
8. *The Private Diary of Andrew Muir: 9 November, 1848 - 5 August, 1850*, Transcript. PABC.
9. D.B. Smith, ed., *The Reminiscences of Doctor John Sebastian Helmcken* (Vancouver: University of British Columbia Press, 1975). p. 306.
10. W. Kaye Lamb, "The Governorship of Richard Blanshard," *British Columbia Historical Quarterly*, XIV. p. 8.
11. T.S. Ashton and J. Sykes, *The Coal Industry of the Eighteenth Century*, 2nd. ed. (Manchester: Manchester University Press, 1964), pp. 70-83.
12. A. Campbell, "Coal Miners: Skill, Independence, and Trade Unionism in the Coalfields of Nineteenth Century Britain, with particular reference to Scotland," *Canadian Historical Association Historical Papers*, 1981. p. 161. Quoting T. Stewart, *Among the Miners* (Larkhall,1893), pp. 17-18.
13. Blanshard to Grey, 8 April 1850.
14. Smith, p. 308.
15. Lamb, p. 9. Quoting Douglas to Simpson, 15 October 1850.
16. Smith, p. 309.
17. *Ibid.* p. 311.
18. Kemble, p. 130
19. Blanshard to Grey, 10 June 1851.
20. J.E. Hendrickson, "Two Letters from Walter Colquhoun Grant," *British Columbia Studies*, No. 26. Summer, 1975. p. 12.
21. *Ibid.*

Chapter III

1. B.A. McKelvie, "The Founding of Nanaimo," *British Columbia Historical Quarterly*, July, 1944. pp. 171-72. Quoting Joseph William McKay, *Recollections of a Chief Trader in the Hudson's Bay Company*.
2. *Ibid.* Quoting Douglas to Barclay, 23 June,1852.
3. *Ibid.* Quoting Douglas to Barclay, 18 August,1852.
4. Douglas to McKay, 24 August, 1852. (The correspondence which passed between Douglas and McKay about the coal find in the Nanaimo area is collected under the title *Nanaimo Correspondence, 1852-53*). PABC.
5. *Ibid.*
6. *Ibid.* Douglas to McKay, 26 August 1852.
7. *Ibid.*
8. *Ibid.* Douglas to McKay, 31 August 1852.
9. *Ibid.* McKay to Douglas, 9 September 1852.
10. *Ibid.* McKay to Douglas, 10 September 1852.
11. *Ibid.* McKay to Douglas, 16 September 1852.
12. *Ibid.* Douglas to McKay, 13 October 1852

13. *Ibid.* Douglas to McKay, 27 October 1852.
14. *Ibid.* Douglas to McKay, 26 August 1852.
15. *Ibid.* McKay to Douglas, 18 May 1853.
16. *Ibid.* Douglas to McKay, 20 May 1853.
17. Ibid McKay to Douglas, 2 June 1852.
18. *Ibid.* Douglas to McKay, 4 June 1852.
19. *Ibid.* Douglas to McKay, 27 September 1853.
20. *Ibid.* McKay to Douglas, 17 July 1853.
21. *Ibid.* Douglas to McKay, 27 September 1853.
22. McKelvie, p. 184.
23. Douglas to McKay, 12 September 1853.
24. Douglas to McKay, 27 September 1853
25. C. Hillier, *A Journey to the Heart of England* (London: Granada Publishing, 1978), p. 67.
26. E. Lipson, *The Growth of English Society* (London: A.& C.Black, 1951), p. 242-3.
27. E.R. Pike, *Human Documents of the Victorian Golden Age* (London: Unwin Ltd., 1967), p. 82.
28. E.B. Norcross (ed.), *Nanaimo Retrospective: The First Century* (Nanaimo: Nanaimo Historical Society, 1979), p. 154.
29. Indenture of John Thompson. PABC.
30. Douglas to McKay, 20 May 1853.
31. McKay to Douglas, 18 May 1853
32. B. Goult, "First and Last Days of the *Princess Royal*", *British Columbia Historical Quarterly*, January 1939, p. 16.
33. *Ibid.* p. 17. Quoting log of the *Princess Royal* 1854-55, kept by First Mate Charles Gale.
34. *Ibid.* pp. 18-19. Quoting log of *Princess Royal*.
35. *Ibid.* pp. 19-20. Quoting letter Douglas to Hudson's Bay Company, London. 23 November 1854.
36 *Ibid.* p. 20. Quoting letter Douglas to Hudson's Bay Company, London. 28 November 1854.
37. McKelvie, p. 187.
38. "The Census of Vancouver Island 1855," *British Columbia Historical Quarterly*, January, 1940. pp. 51-55. (Date of census 31 December 1854).
39. M.A. Ormsby, *Britsh Columbia: A History* (Vancouver: McMillan, 1958), p. 115.
40. *Fort Nanaimo Journal*, 25 August 1855. PABC.
41. McFie, p.145.
42. *Ibid.* pp. 146-8. Quoting Director's Report, Vancouver Coal Mining and Land Company, 29 November 1864.

Chapter IV

1. D. Pethick, *Summer of Promise: Victoria 1864-1914* (Victoria: So No Nis Press, 1980), p. 115.
2. J. Scott, *Sweat and Struggle* (Vancouver: New Star Books, 1974), p. 155.
3. J. Audain, *Coalmine to Castle* (Victoria: Skylane Publishing Co., 1964), p. 7.

4. M. Wilson, *The Ayrshire Hermit - Tammie Raeburn, Hurlford Sixty Years Ago, Its Rise, Condition, and Celebrities, Including Davies Clark the Centarian* (Kilmarnock: Alfred Chas. Jonas, 1875), pp. 64-64.
5. Ashton and Sykes, Ashton, pp. 1-2.
6. Scott, p. 150.
7. B.E. Paterson, *The Social and Working Conditions of the Ayrshire Mining Population 1840-1875* (Ayrshire Archeological and Natural History Society, 1972), pp. 211-214.
8. Kilmarnock Journal, 6 June 1850.
9. Ashton and Sykes, pp.1-2.
10. W.E. Houghton, *The Victorian Frame of Mind: 1830-1870* (New York: Yale University Press, 1957), p. 185.
11. *Ibid.*
12. G. Myers, *A History of Canadian Wealth* (Chicago: Kerr, 1914), p. xxxv.
13. Copy on display at Craigdarroch Castle, Victoria.
14. Kilmarnock Journal, 5 December 1850.
15. Kilmarnock Journal, 19 December 1850.
16. Douglas to McKay, 20 May 1853.
17. *Nanaimo Journal*, 12 October 1855.
18. Norcross, p. 55.
19. Audain, pp. 36-38.
20. Audain, p. 44.
21. Nanaimo Free Press, 30 March 1866.
22. Audain, pp. 49-50.
23. Audain, quoting a letter from R. Dunsmuir to H.H. Langevin, Minister of Public Works, British Columbia, n.d. p. 50.
24. Articles of Partnership: Dunsmuir, Diggle & Co., PABC.
25. Indenture: Land Rental. McCallum/Dunsmuir etc. 21 December 1871. PABC.
26. Audain, pp. 52-53.
27. D.T. Gallacher, *Dictionary of Canadian Biography* (Toronto: University of Toronto Press, 1982), XI, p. 291.
28. Audain, p. 65.
29. Colonist, 14 September 1883
30. Dissolution of Partnership: Dunsmuir, Diggle & Co., May 12, 1883. PABC.
31. Colonist, 24 April 1877.
32. P. Berton, *The National Dream* (Toronto: McClelland and Stewart, 1970. pp. 188-189).
33. Mainland Guardian, February 3,1877.
34. M. Robin, *The Rush for Spoils: The Company Province 1871- 1933* (Toronto: McLelland and Stewart, 1972), p. 31.
35. Myers, pp. 302-304.
36. L. Bowen, *Boss Whistle* (Lantzville: Oolichan Books, 1982), pp. 44-45.
37. Reprint, Colonist, 30 June 1918. Vert. File. PABC.
38. British Columbia Legislative Journals, 1888.
39. Scott, p. 182.

40. Colonist, 13 April 1889.
41. *Ibid.*
42. Colonist, 17 April 1899.
43. Colonist, 10 January 1899.
44. Robert Dunsmuir Will. PABC.

Chapter V

1. Nanaimo Free Press, 13 March 1875.
2. *Fort Nanaimo Journal*, 12 October 1855.
3. Nanaimo Free Press, 13 February 1877.
4. Nanaimo Free Press, 28 February 1877.
5. Lipson, p. 248.
6. 1902 Royal Commission to Investigate Chinese and Japanese Immigration into British Columbia, p. 78.
7. Nanaimo Free Press, 7 and 14 March 1877.
8. Colonist, 13 and 14 March 1877.
9. Nanaimo Free Press, 7 March 1877.
10. *Ibid.*
11. Colonist, 17 March 1877.
12. Nanaimo Free Press, 28 February 1877.
13. Colonist, 1 March 1877.
14. Nanaimo Free Press, 3 March 1877.
15. Colonist, 29 May 1877.
16. Colonist, 8 April 1877.
17. *Ibid.*
18. Colonist, 15 April 1877.
19. Colonist, 17 March 1877.
20. *Ibid.*
21. Nanaimo Free Press, 14 March 1877.
22. Colonist, 17 March 1877.
23. Colonist, 8 April 1877.
24. *Ibid.*
25. *Ibid.*
26. Colonist, 25 February 1877.
27. Colonist, 24 March 1877.
28. Lt.Col.Houghton, Letterbook I: Correspondence Out, 26 April 1877. PABC.
29. Scott, p. 171.
30. Military Requisition to Lt.Col.Houghton, April 28, 1877.PABC.
31. Letterbook I. Houghton to Adjutant General, April 28, 1877
32. Mainland Guardian, 2 May 1877.
33. Mainland Guardian, 4 May 1877.
34. Letterbook I, Military Report.

35. Mainland Guardian, 4 May 1877.
36. Colonist, 3 May 1877.
37. *Ibid.*
38. Letterbook I, Military Report.
39. *Ibid.*
40. Colonist, 2 May 1877.
41. Mainland Guardian, 5 May 1877.
42. Mainland Guardian, 10 May 1877.
43. Colonist, 12 May 1877.
44. Colonist, 13 May 1877.
45. Colonist, 12 May 1877.
46. Colonist, 13 May 1877.
47. Colonist, 27 May 1877.
48. Colonist, 29 May 1877.
49. *Ibid.*
50. Colonist, 9 June 1877.
51. Colonist, 10 January 1899.

Chapter VI

1. Colonist, 29 March 1877.
2. Colonist, 22 March 1877.
3. Colonist, 29 March 1877
4. B.C. Sessional Papers, 1877. p. 504.
5. Letterbook: B.C.Inspector of Mines. Letters ff. November 1, 1877 - August 15, 1878.
6 Nanaimo Free Press, 4 May 1887.
7. Nanaimo Free Press, 11 May 1887.
8. Nanaimo Free Press, 7 May 1887.
9. Colonist, 18 May 1887.
10. Colonist, 26 June 1887.
11. Colonist, 18 May 1887.
12. B.C. Sessional Papers, 1888. pp. 367 - 379.
13. Colonist, 1 June 1887.
14. Letterbook: Inspector of Mines. Examination of No.5 Pit, Wellington, July 19, 1887.
15. Letterbook. Dick to Robson, January 27, 1888.
16. Colonist, 25 January 1888.
17. *Ibid.*
18. Colonist, 27 January 1888.
19. Colonist, 25 January 1888.
20. *Ibid.*
21. Colonist, 4 February 1888.
22. B.C. Sessional Papers, 1889. p. 338.
23. *Ibid.* p. 339.

24. 1902 Commission on Chinese and Japanese Immigration into British Columbia. pp. 76 - 79.
25. Colonist, 5 February 1888.
26. Colonist, 7 February 1888.
27. B.C. Legislative Journal, 1888. April 5.
28. Colonist, 9 March 1888.
29. J. Morton, *In a Sea of Sterile Mountains* (Vancouver: J.J. Douglas Ltd., 1974), p. 161.
30. B.C. Sessional Papers, 1899. p. 1186.
31. B.C. Sessional Papers, 1892. pp. 465-475.

Chapter VII

1. Robert E. Wynne, *Reaction to the Chinese in the Pacific Northwest and British Columbia, 1850-1910* (New York: Arno Press, 1978, pp. 6 - 7.
2. *Ibid.* p. 25. Quoting R. Connell, *Why and How . . .* (Boston, 1871), p.125.
3. *Ibid.* p. 4. Quoting H.H. Bancroft, *Retrospectives* (New York: 1912), p. 353.
4. Victoria Gazette, 30 June 1859.
5. Colonist, 10 May 1860.
6. Douglas to Newcastle. 23 April 1860.
7. Morton, p. 10.
8. Colonist, 5 March 1860.
9. Colonist, 6 March 1860.
10. McFie, pp. 382 - 388.
11. Colonist, 8 May 1867.
12. Colonist, 27 January 1871.
13. 1885 Royal Commission Report on Chinese Immigration into British Columbia. p. 118.
14. Colonist, 7 May 1871
15. Colonist, 2 September 1871.
16. W. Peter Ward, *White Canada Forever* (Montreal: McGill- Queen's University Press, 1978, Table 1. p. 170.
17. Report of the Inspector of Mines, 1881.
18. Morton, p. 50
19. Colonist, 2 May 1876.
20. 1885 Report, p. iv.
21. *Ibid.* p. ix.
22. *Ibid.* p. cii.
23. *Ibid.* Evidence: J. Metcalfe, pp. 82 - 84; J. Young, pp. 87 - 90; R. Scott, pp. 107 - 108; J. Bryden, pp. 110 - 112; S. Robins, pp. 118 - 119; R. Dunsmuir, pp. 127 - 131; J. Pawson, pp. 131 - 134; Knights of Labor, pp. 155 - 160; Nanaimo Trades Association, pp. 196 - 170.
24. Ward, pp. 40 - 41.
25. Nanaimo Free Press, 5 February 1887.
26. S. Shimuzu, Japanese Consul, Vancouver to Earl of Aberdeen, Governor-General of Canada, 10 May 1898.

27. Prime Minister Chamberlain to Earl of Aberdeen, July 20, 1898.
28. Kato, Japanese Legation, to Marquis of Salisbury, August 3, 1898.
29. 1902 Royal Commission Report on Chinese and Japanese Immigration into British Columbia. Evidence: S. Robins, pp. 72 - 76; F.D. Little, pp. 76 - 79; J. Matthews, pp. 79 - 80; A. Brydon, pp. 80 - 81; W. Woodman, p. 82; Nanaimo Miners Union, pp. 83 - 85; Alexandra Miners Union, pp. 85 - 86.

Chapter VIII

1. Colonist, 9 July 1890.
2. Colonist, 24 July 1890.
3. J. Audain, *Alexander Dunsmuir's Dilemma* (Victoria: Sunnylane Publishing Co., 1964), ff. 3-31.
4. Colonist, 30 July 1890.
5. *Ibid.*
6. Colonist, 1 August 1890
7. Colonist 29 July 1890.
8. *Ibid.*
9. Colonist, 5 August 1890.
10. Colonist, 7 August 1890.
11. British Columbia Legislative Journals, 1891. "Report of the Select Committee: Wellington Strike."
12. *Ibid.*
13. Colonist, 13 July 1890
14. Colonist, 31 July 1890.
15. Colonist, 6 August 1890.
16. British Columbia Legislative Journals, 1891. Correspondence: Sending Militia to Wellington.
17. Colonist, 10 August 1890.
18. Scott, pp. 196 - 198.
19. *Ibid.*, p. 104. Quoting Palladium of Labour, January 20, 1885.
20. Colonist, 6 March 1890.
21. Colonist, 8 March 1891.
22. British Columbia Legislative Journals, 1891. Report of Select Committee: Attack on Ellis Roberts Funeral.
23. Colonist, 11 March 1891.
24. Colonist, 2 April 1891.
25. Canadian Royal Commission Report: Commission Relating to Unrest and Discontent Among Miners and Mine-owners in the Province of British Columbia, 1899. pp. 283 - 285.

Chapter IX

1. Nanaimo Free Press, 14 June 1876.
2. Colonist, 2 April 1891.

3. J.Cass, *The Wellington Story 1869 - 1900.* pp. 1 - 10. Mimeo. Vert. File. PABC.
4. Bowen, pp. 24 - 25.
5. British Columbia Legislative Journals, 1885. p. 25.
6. Chuen-Yan David Lai, "Home County and Clan Origins of Overseas Chinese in Canada in the Early 1880's," *British Columbia Studies*, No. 27, Fall 1975.
7. Ward, p. 3. 8. Colonist, 18 May 1888.
9. Norcross, p. 131.
10. J. Cass, "Notes on Nanaimo Pioneers", Vert. File. PABC.
11. *Ibid.*
12. Audain, Dilemma. pp. 117 - 120.

Bibliography

Ashton, T.S. and J. Sykes, *The Coal Industry of the Eighteenth Century*, 2nd. ed. (Manchester: Manchester University Press, 1964).

Audain, *J. Alexander Dunsmuir's Dilemma* (Victoria: Sunnylane Publishing Co., 1964).

Audain, J. *Coalmine to Castle* (Victoria: Skylane Publishing Co., 1964).

Bancroft, H.H. *History of British Columbia* (San Francisco: 1887).

Berton, P. *The National Dream* (Toronto: McClelland and Stewart, 1970).

Bowen, L. *Boss Whistle* (Lantzville: Oolichan Books, 1982).

Cass, J. *The Wellington Story, 1869 - 1900*. Mimeo. Vert. File. PABC.

Gough, Barry M. *The Royal Navy and the North West Coast of America: 1810 - 1914* (Vancouver: University of Vancouver Press, 1971).

Guillet, E.C. *The Great Migration*, 2nd. ed. (Toronto: Toronto University Press, 1963).

Hillier, C. *A Journey to the Heart of England* (London: Granada Publishing, 1978).

Holbrook, Stewart H. *The Columbia* (Toronto: Rinehart and Co. Inc., 1956).

Houghton, W.E. *The Victorian Frame of Mind: 1830 - 1870* (New York: Yale University Press, 1957).

Howay, F.W. *British Columbia: The Making of a Province* (Toronto: Ryerson Press, 1928).

Leynard, A. *The Coal Mines of Nanaimo*. Paper presented to the Nanaimo Historical Society, April 20, 1892 Lipson, E. *The Growth of English Society* (London: A.& C. Black, 1951).

MacFie, M. *Vancouver Island and British Columbia* (London: Longman, Green, Longman, Roberts and Green, 1865).

Morton, A.S. *The Canadian West to 1870 - 71*, 2nd. ed. (Toronto: University of Toronto Press, 1973).

Morton, J. *In a Sea of Sterile Mountains* (Vancouver: J.J. Douglas Ltd., 1974).

Myers, G. *A History of Canadian Wealth* (Chicago: Kerr, 1914).

Norcross, E.B. (ed), *Nanaimo Retrospective: The First Century* (Nanaimo: Nanaimo Historical Society, 1979).

Paterson, B.E. *The Social and Working Conditions of the Ayrshire Mining Population 1840 - 1875* (Ayrshire Archeological and Natural History Society, 1972).

Ormsby, M.A. *British Columbia: A History* (Vancouver: McMillan, 1958).

Pethick, D. *Summer of Promise: Victoria 1864 - 1914* (Victoria: Sono Nis Press, 1980).

Pike, E.R. *Human Documents of the Victorian Golden Age* (London: Unwin Ltd., 1967).

Robin, M. *The Rush for Spoils: The Company Province 1871 - 1933* (Toronto: McLelland and Stewart, 1972).

Scott, J. *Sweat and Struggle* (Vancouver: New Star Books, 1974).

Smith, D.B. ed., *The Reminiscences of Doctor John Sebastian Helmcken* (Vancouver: University of British Columbia Press, 1975).

Tolmie, William F. *The Journals of William Fraser Tolmie: Physician and Fur Trader* (Vancouver: Mitchell Press, 1963).

Ward, W. Peter *White Canada Forever* (Montreal: McGill- Queen's University Press, 1978.

Wilson, M. *The Ayrshire Hermit — Tammie Raeburn, Hurlford Sixty Years Ago, Its Rise, Condition, and Celebrities, Including Davies Clark, the Centarian* (Kilmarnock: Alfred Chas. Jonas, 1875).

Wynne, Robert E. *Reaction to the Chinese in the Pacific Northwest and British Columbia, 1850 - 1910* (New York: Arno Press, 1978).

Articles

Anon., "The Census of Vancouver Island, 1855," *British Columbia Historical Quarterly*, January, 1940.

Anon., "The Founding of Victoria," *Beaver*, March, 1943.

Campbell, A. "Coal Miners: Skill, Independence, and Trade Unionism in the Coalfields of Nineteenth Century Britain, with particular reference to Scotland," *Canadian Historical Association Historical Papers*, 1981.

Chuen-Yan, David Lai. "Home County and Clan Origins of Overseas Chinese in Canada in the Early 1880's, *British Columbia Studies*, No. 27, Fall, 1975.

Goult, B. "First and Last Days of the Princess Royal", *British Columbia Historical Quarterly*, January, 1939.

Hendrickson, J.E. "Two Letters from Walter Colquhoun Grant," *British Columbia Studies*, No. 26, Summer, 1975.

Ireland, W.E. "The Appointment of Governor Blanshard, *British Columbia Historical Quarterly*, July, 1944.

Kemble, J.H. "Coal from the North West Coast 1848 - 1850," *British Columbia Historical Quarterly*, April, 1938.

Lamb, W. Kaye. "The Governorship of Richard Blanchard," *British Columbia Historical Quarterly*, XIV, 1964.

McKelvie, B.A. "The Founding of Nanaimo," *British Columbia Historical Quarterly*, July, 1944.

Newspapers

Colonist, Victoria, British Columbia.
Kilmarnock Journal, Kilmarnock, Scotland.
Mainland Guardian, New Westminster, British Columbia.
Nanaimo Free Press, Nanaimo, British Columbia.
Nanaimo Journal, Nanaimo, British Columbia.
Victoria Gazette, Victoria, British Columbia.

Documents

(Chronologically arranged)

Richard Blanshard, Vancouver Island Governor, Correspondence Outward 1849 - 1851. PABC.

The Private Diary of Andrew Muir 9 November 1848 - 5 August 1850, Transcript. PABC.

Marriage Certificate of Robert and Joan Dunsmuir. Copy on display at Craigdarroch Castle, Victoria.

Nanaimo Correspondence, 1852 - 1855. Collected letters passed between Douglas and McKay regarding coal discoveries at Nanaimo. PABC.

Fort Nanaimo Journal. PABC.

Indenture: John Thompson, coalminer, and Hudson's Bay Company. PABC.

Articles of Partnership: Dunsmuir, Diggle & Co., PABC.

Lt. Col. Houghton. Letterbook I: Correspondence Out. PABC.

Military Requisiton to Lt. Col. Houghton. April 28, 1877. PABC.

Letterbook I: Military Report, 1877. PABC

British Columbia Sessional Papers, 1877.

Letterbook: British Columbia Inspector of Mines.

Dissolution of Partnership: Dunsmuir, Diggle and Co., May 12, 1883. PABC.

1885 Royal Commission Report on Chinese Immigration into British Columbia.

British Columbia Legislative Journal, 1888.

British Columbia Sessional Papers, 1888 and 1889.

British Columbia Legislative Journals, 1891."Report of the Select Committee: Wellington Strike."

British Columbia Legislative Journals, 1891. "Report of Select Committee: Attack on Ellis Roberts Funeral."

British Columbia Legislative Journals, 1891. "Correspondence: Sending Militia to Wellington."

British Columbia Sessional Papers, 1892.

S. Shimuzu, Japanese Consul, Vancouver to Earl of Aberdeen, Governor General of Canada, May 10, 1898.

Prime Minister Chamberlain to Earl of Aberdeen, July 20, 1898.

Kato, Japanese Legation, to Marquis of Salisbury, August 3, 1898.

British Columbia Sessional Papers, 1899.

Canadian Royal Commission Report: Commission Relating to Unrest and Discontent Among Miners and Mine-owners in the Province of British Columbia, 1899.

1902 Royal Commission to Investigate Chinese and Japanese Immigration into British Columbia.

Index

About the author

Eric Newsome was born in 1925 in a Yorkshire coal mining village where his family — father, grandfathers, and uncles — all went underground to earn their living. When he left school at fourteen, he went to work in a shipyard; anything was better than going down into the mines.

During World War II, Newsome was first an air cadet and then an R.A.F. pilot from 1943 - 48. After the war he bluffed his way into teacher training and taught in schools in England and in the British Zone in Germany before emigrating to Canada in 1958. He taught in Montreal while earning his degrees (B.A., Sir George Williams; M.Ed., Bishop's University) the hard way, at night and during the summers. His love of flying led Newsome to take up gliding, which he did seriously enough to compete at the Canadian National Soaring Championships, and to be Chief Flying Instructor at clubs both in Montreal and Vancouver.

Eric Newsome, tired of trying to please too many masters as an administrator in "a giant education factory," took early retirement in 1979, and headed west to British Columbia to write. He lives with his wife Irene in Abbotsford and teaches Geography in Vancouver area colleges ("to fight off brain-rust and because, after forty years, I don't know how to stop"). And he writes every day.

Mr. Newsome's second book, *The Case of the Beryl G*, a true story of hijacking and murder on the west coast during the wild era of American prohibition, will be published by Orca Book Publishers in the fall of 1989.